# OUR DAILY BREAD

# OUR DAILY BREAD

## *The Peasant Question and Family Farming in the Colombian Andes*

NOLA REINHARDT

UNIVERSITY OF CALIFORNIA PRESS
BERKELEY LOS ANGELES LONDON

University of California Press
Berkeley and Los Angeles, California

University of California Press, Ltd.
London, England



Library of Congress Cataloging-in-Publication Data
Reinhardt, Nola.
Our daily bread : the peasant question and family farming in the Colombian Andes / Nola Reinhardt.
p. cm.
Bibliography: p.
Includes index.
ISBN 0-520-06225-6 (alk. paper)
1. Family farms—Colombia. 2. Peasantry—Colombia. I. Title.
HD1476.C7R45 1989
338.1'09861—dc19 88-14370
CIP

Printed in the United States of America

1 2 3 4 5 6 7 8 9

# Contents

# Tables, Figures, and Maps

## Tables

## Figures



## Maps

# Acknowledgments

I would like to acknowledge my debt of gratitude for the institutional and individual support that I received in the various stages of this work, which was begun as a Ph.D. thesis in economics at the University of California at Berkeley. The Center for Latin American Studies at Berkeley and the Tinker Foundation provided a travel grant for my first visit to Colombia in the spring of 1978, when my preliminary conceptions were tested against the reality of Colombian agriculture and the desirability of an in-depth study of mountain agriculture became clear. I am grateful to Richard Sutch and Leonard Joy for their extremely valuable assistance during the next stage of preparation for the fieldwork I had decided to undertake—including essential advice on topics ranging from the mechanics of designing the interview form to the subtleties of the interview process itself.

The Inter-American Foundation funded a year-long residence in Colombia, from September 1978 to July 1979, during which I consulted various sources in Cali, Bogotá, and Pasto and carried out the detailed study of El Palmar. I would like to thank the institutions in Colombia whose facilities were so graciously made available to me on that trip: the Universidad del Valle in Cali, the Corporación Autónoma Regional del Valle del Cauca in Cali (with special thanks to the librarians), the Federación Nacional de Cafeteros Colombianos in Cali and Pasto, the Comité Regional de Producción Agrícola del Valle del Cauca in Cali, the Instituto Geográfico Agustín Codazzi in Cali and Dagua, the Departamento Administrativo Nacional de Estadística in Cali and Bogotá, and the Instituto Colombiano Agropecuario in Pasto. The following individuals also helped me substantially: Alberto and Ángela Corchuelo, Nelly Acevedo, Gonzalo Pereáñez, Jairo Roa Narvaez, Diego Roldán Luna, Gustavo de Roux, and Fernando Uribe.

A Picker Fellowship from Smith College in 1983 enabled me to return to Colombia from December 1983 through January 1984 to extend the original research both historically and regionally and to update the study of El Palmar. I would like to thank the individuals and institutions in Valle whose assistance was so helpful on this second visit: Silva Holguin and the director of the Archivos Históricos in Cali, the Biblioteca Departamental in Cali, José and Blanca Escorcia, Fernando Urrea Giraldo, the librarians in the census division of the Departamento Administrativo Nacional de Estadística in Cali, the parish of Dagua, and the many individuals and institutions in Cali who supplemented their earlier assistance. The generous scholarship received from the Picker Fund provided me with funds and time to write this book upon my return from Colombia in 1984. I would like to thank the student assistants who worked with me under the Picker grant: Rosemary Reed, Leigh Peake, Margaret Kane, Laurel Touby, Kathy O'Brien, and Pamela Wood. Their willing help in carrying out many of the frequently tedious aspects of the project was greatly appreciated.

Many people in the United States have provided valuable insights and helpful comments during various stages of this project, and I am grateful for their advice, while absolving them of any responsibility for the final results: Charles Bergquist, Claudia Carr, Carmen Diana Deere, Alain de Janvry, Arturo Escobar, Albert Fishlow, William Glade, Joan Hannon, Richard Hyland, Catherine LeGrand, James Parsons, Michael Redclift, Brian Sheppard, Michael Taussig, William Thiesenhusen, Richard Walker, Benjamin Ward, Michael Watts, Andrew Zimbalist, and the staff and fellow Research Fellows of the Inter-American Foundation. Members of the New England Women in Development group provided useful comments on the first draft of this work, and I owe them a particular debt of thanks. My editors at the University of California Press, Naomi Schneider and Barbara Ras, were most helpful and encouraging throughout the various stages of getting the book to press.

Most of all, I would like to thank the people without whom this book would not have been possible: the women, men, and children of El Palmar who gave so generously of their time and their hospitality. I trust that they will find this book to be faithful to their histories as they have recounted them to me. Finally, my undying gratitude and respect go to Diomelina Álvarez de Benavides and her family, whose spirit illuminates this book and provides a shining hope for the future of Colombia.

# Introduction

This book is about the "peasant question": the impact of economic development on peasant agriculture. At one level, it is a case study of a community of smallholding producers in the Colombian Andes, a study that illustrates the general pattern of Colombian mountain settlement in the nineteenth century and the complex process of development of those communities during the twentieth century. At a second level, it is a study in regional economic development. At a third level, it is an analysis of Colombian agricultural policy and of the emerging structure of Colombian agriculture. Finally, and at a more fundamental level, it is the application of a methodology and a theoretical framework to analyze the changing status of peasant production (or, as this form of agricultural production is also often called, family farming).[1]

The "peasant question" has been a matter of debate for the past decade in many parts of the world, including Latin America. Much of the literature on Latin American agriculture up to the 1970s presented a picture of a dual agrarian structure with large estates (*latifundios*) controlling most of the land, while the majority of agricultural units were subsistence or below-subsistence operations (*minifundios*) related to the estates through traditional land and labor arrangements.[2] The latifundios generally engaged in extensive

1. "Peasant production" and "family farming" will be used interchangeably here to denote predominantly family-labor-based agricultural production, for home consumption or market exchange, carried out for purposes of household sustenance.

2. Classic and influential examples of this literature are Feder 1971 and the

cattle-raising activities combined with tenant peasant production of foodstuffs. By the 1970s, however, estate production in many Latin American countries was undergoing a transition from traditional agricultural activities to modern, capital-intensive, commercial agricultural production. As this transition occurred, what was happening to the peasant holdings?

Some writers believed that this process of estate modernization was weakening and eliminating small-scale agricultural production. With respect to Colombia, Kalmanovitz argued this forcefully in an important study for the Colombian national statistics bureau (División Administrativa Nacional de Estadística, or DANE) published in 1974. The most recent proponent of this position for Latin America as a whole is de Janvry (1981). On the other hand, a number of writers argued on either theoretical or empirical grounds that small-scale agricultural production in Latin America continued to be important: for example, Esteva (1978) and, more recently, Goodman and Redclift (1982).[3]

The present study examines both the theoretical and empirical aspects of the "peasant question" in Latin America. Many writers arguing for the persistence of peasant production believe that certain aspects of peasant decision-making behavior (for example, the willingness of the farm household to forgo rent, profits, and sometimes even the opportunity cost of labor) give family farming an inevitable economic superiority over large-scale production. This general theoretical position is frequently referred to as "Chayanovian." The classic Marxist position, on the other hand, argues for the economic superiority of large-scale capitalist agriculture based on its advanced level of technological development. I review these two perspectives in chapter 1 and argue that the impact of economic development on agrarian structure cannot be as definitively

---

CIDA-sponsored studies of agrarian structure carried out in seven Latin American countries in the early 1960s, the results of which are summarized in Barraclough and Domike 1966 and Barraclough 1973.

3. On this debate, see also the three issues of *Latin American Perspectives* on the peasantry, nos. 18 (Summer 1978), 19 (Fall 1978), and 27 (Fall 1980); the April 1982 issue of the *Revista de la Cepal* (no. 16); and in Colombia, Corchuelo 1979; Gómez Villa 1981; Zamosc 1979; and Moncayo and Rojas 1979. An intermediate position is taken by Roseberry in a recent work (1983), where he shows how peasant production in Peru has been both positively and negatively affected by the economic transformation of agricultural production.

predicted as they suggest. On an empirical level, I show that the Latin American agrarian sector is currently characterized by the coexistence of peasant and capitalist agriculture, rather than by the dominance of one or the other form of production.

The basis of this coexistence is the focus of this study. Its underlying theoretical perspective is that the emergence, persistence, or elimination of family farming is not inevitable, but rather problematic. The methodological conclusion that follows is that the "peasant question" must be examined in each specific case as the outcome of a concrete, historical, social process, which either assures or undermines the conditions of existence of this organizational form. The analysis needs to encompass both the specific economic functioning of the family farm ("internal" factors) and the constellation of broader social forces operating on it ("external" factors). This is the approach that is applied here to the study of the "peasant question" in the Dagua region of southwestern Colombia.

Colombian agriculture today is characterized by a diverse regional structure: large-scale cotton cultivation in the northern coastal plains; the primarily cattle-raising eastern plains (*llanos*), with scattered zones of recent small-scale colonization; the highly commercial large-scale agriculture of the interior river valleys, growing such crops as sugar cane (for refined sugar), sesame, soybeans, and other feedgrains; the coffee estates of the middle and western mountain zones; and, finally, the food- and coffee-growing family farms scattered throughout Colombia's three mountain ranges. The three largest cities of Colombia—Bogotá, Medellín, and Cali—are surrounded by regions of small-scale mountain agriculture that provide the metropolitan zones with the bulk of such foods as maize, kidney beans, *panela* (unrefined block sugar), tomatoes, string beans, herbs, plantains, onions, potatoes, cassava (manioc), and *arracacha* (a starchy tuber).[4]

4. An excellent study of peasant agriculture in the region of Bogotá was recently carried out by a research team from the Pontificia Universidad Javeriana (Forero and Rudas 1983); a detailed study of one peasant community in that area, modeled after this study of El Palmar in Dagua, is currently being carried out by Humberto Rojas Ruiz. An older study of a peasant community in that region is Fals Borda 1957; see also Fals Borda 1955. A number of detailed studies of peasant agriculture in the mountain regions surrounding Cali have been carried out in recent years by the Corporación Regional Autónoma del Valle del Cauca (CVC), and include CVC 1971, 1976, 1977a and Leaño Castellanos 1973. The Programa de

I selected the Dagua region in the mountains to the west of Cali for this study of peasant production after extensive research had established it as a representative zone of food-producing small-scale agriculturalists.[5] Dagua is a county (*municipio*) in Valle del Cauca, the department containing the city of Cali and the highly developed commercial agriculture of the Cauca River Valley.[6] A region of diversified food and coffee cultivation, Dagua is of varied elevation, which creates a number of distinct agricultural zones.[7] The coffee zones of Dagua are marginal to the main Colombian coffee-producing region to the north: overall, Dagua is among neither the richest nor the poorest mountain regions of the country. The distribution of agricultural units in Dagua in 1970 was comparable to that in the nation as a whole, with units of less than ten hectares (one hectare equals 2.4 acres) comprising just over 70 percent of agricultural units in both cases (DANE 1973a).[8]

---

Desarrollo Rural Integrado (DRI, the government's integrated rural development program) has carried out a number of studies of smallholding agriculture around the country and is currently studying the Dagua region. Other published studies of family farming in Colombia include Parsons 1963 on Antioquia (which includes the Medellín region); Havens 1966; Haney 1969; Pearse and Rivera 1963; Sandoval 1966; Smith 1944; Instituto Colombiano Agropecuario (ICA) 1972, 1973; Inter-American Development Bank (IDB) 1972; and PRODESARROLLO 1968, 1972b. Unpublished studies have also been carried out by other groups, including the Federación Nacional de Cafeteros Colombianos (the Colombian coffee-growers association) and the Centro Internacional de Agricultura Tropical (CIAT, an international agronomic research institute). The personnel of agencies such as the CVC, the DRI, and the Federación de Cafeteros are invaluable sources of information about mountain agriculture, as are researchers at the nation's universities.

5. In addition to the studies mentioned above and the Colombian agricultural censuses of 1960 and 1970, I consulted with researchers and field representatives of a number of agencies in Colombia, including the CVC in Cali, the Instituto Geográfico Agustín Codazzi (the Colombian geographic institute, responsible for carrying out land surveys and maintaining cadastral records) in Cali and Dagua, the Federación de Cafeteros in Cali, Bogotá, Dagua, Nariño and Huila, and the DRI in Bogotá, as well as with researchers at the Universidad del Valle.

6. The department (*departamento*) is the highest-level political division in Colombia, similar to the state in the United States, but with less autonomy from the central government (the president of Colombia appoints the governors of each *departamento*).

7. The commercial products of the Dagua region, in addition to coffee, include *panela*, plantains, maize, beef and beef products, pork, eggs, cassava, *arracacha*, kidney beans, a variety of fruits and herbs, string beans, tomatoes, green peppers, and honey.

8. The distribution of land ownership in Dagua is less concentrated, however,

Dagua is traversed by a highway connecting Cali to the major Pacific Coast port of Buenaventura, which provides the growers of the region with varying degrees of access to the valley markets. A new route for part of this highway was completed in 1968, passing through the *vereda* (community) of El Palmar. I chose this community for in-depth study during the 1978–79 agricultural year, again after consultation with local experts.

El Palmar was selected for a number of reasons. First, because it had been settled in the late 1800s and early 1900s, a number of the early settlers were still alive. As there are few historical documents with information on the small-scale agriculturalists of Colombia (except for some parish registries and some records of legal battles with large landowners), oral histories are of vital significance. Much of the history of the settlement of the Dagua region that I present is based on such oral histories.[9]

Second, in many respects the community was typical of smallholding agricultural communities in the region. The size-distribution of holdings was comparable to that in other mountain communities already studied by the Corporación Regional Autónoma del Valle del Cauca (CVC), an important agronomic agency in Valle. El Palmar lies at an elevation of 1,400 to 1,700 meters: the cultivated land in the mountain zone runs from approximately 1,200 to 2,000 meters. Its soils are of average quality for the zone. Land in the community was used for the production of mixed foodstuffs and coffee, as was common for areas of this elevation in the region.[10]

Although El Palmar was in many respects typical, it was selected also because of a particular atypicality: the fact that the new path of

than for the nation as a whole, with 55.6 percent of the land held by units of over 100 hectares as opposed to 67.5 percent nationally. Some reasons for this pattern are presented in chapter 2.

9. The tapes of these histories have been transcribed in Spanish and can be made available to interested researchers.

10. See note 4 above for CVC studies of Cali-area mountain communities. Mean farm size for the ninety-two landowning households interviewed in El Palmar in 1978–79 was 11.7 acres, while the median was 6 acres (distributional data are presented in chapter 6). The soils of the El Palmar region are derived from igneous rock (decomposition of basalt), with only a superficial depth, moderate resistance to erosion, slow permeability, a pH of 5.5, of "high fertility but low productivity" because of a lack of nitrogen (PRODESARROLLO 1972a).

the Cali-Buenaventura highway passes through it. Even a cursory inspection indicated that the impact of this new transportation route on the economic life of the community had been considerable. In 1978 one could see the beginnings of a small village clinging to the sides of the highway in what had previously been a dispersed farm settlement. A central market sold local products to passersby: beef and pork swinging from huge hooks by the meat stalls, blocks of unrefined sugar, plantains, pineapples, oranges, a scattering of other fruits, and root crops such as cassava and *arracacha.* On weekends the market sold products brought in from Cali to the local residents: potatoes, carrots, onions, cilantro, tomatoes, cabbage, and occasionally other fruits and vegetables. By day the area throbbed with activity, a hum of cars, and conversation, punctuated by the squealing of pigs about to be slaughtered.

Wherever the road gave purchase, somebody had set up a business (*negocio*): a stall or small restaurant selling snacks and quick meals to truckers and travelers, or a small store selling soft drinks, beer, *aguardiente* (a sugar-cane liquor), bread, and penny candy, sometimes with a freezer to keep penny ices, and occasionally stocking a small range of goods for the local residents such as rice, beans, sugar, vegetable oil, potatoes, eggs, kerosene, and coffee. A small stall sold fruit drinks. There was a kiosk with a loudspeaker system for dancing. At night, especially on weekends, the music from this system would compete with the music pouring from the speakers in front of each little store—turned into a bar at dusk—and with the intermittent protesting clamor of trucks and buses struggling up the road from Buenaventura. In a house by the roadside, conversation often stopped while the vehicles passed.

El Palmar was still predominantly agricultural: back off the highway, down one of the paths leading into the countryside, the sounds of the farmland muted the music of the bars and kiosks.[11] Nevertheless, the occasional groan of a passing vehicle could be heard, a reminder that the impact of the highway had extended deeply into the surrounding agricultural community. As the study

11. Of the total net income of $138,125 earned by the 105 resident households interviewed in 1978–79, 85 percent came from agricultural activities: 65 percent from farm production (including the value of produce consumed on-farm), and 20 percent from agricultural wage labor (including earnings from sharecropping arrangements).

of El Palmar reveals, the highway was one of a complex of externally generated changes in the 1970s that affected not only the kinds of agricultural activities being pursued in the community but also the social relations of production and marketing.

Because of its particular atypicality, then, El Palmar offered an excellent opportunity to study the impact of increased economic integration on family farming. In order to understand the effects of recent developments, however, it was necessary to have a baseline from which to gauge the changes. This required an examination of the prior economic and social organization of the community. As I pursued this study, I was drawn back further and further into the history of the community, of the region, and of the nation. Through this research, I sought to uncover the roots of the emergence of the family farm as a form of production in this region and to reconstruct the historical process by which this organizational form was strengthened or threatened as the nation developed. The results provide an example of the way in which economic development can support peasant production, in a manner similar to that recently demonstrated for Guatemala by Smith (1984b) and for Venezuela by Roseberry (1983).

Chapter 2 begins by exploring the colonial patterns of land use established in the region between Cali and Buenaventura, and shows how family farming was gradually extended in an area initially dominated by mining and large estates. Settlers moved into this mountain area in the 1800s from regions south of Valle.[12] These migrants acquired land (although not title) in Dagua and constituted themselves as family units producing a variety of agricultural commodities for local and regional markets and, in some zones, coffee for export. Although there was a wage-labor market, the farms relied heavily on family labor. This labor was organized through a system of internal relations that some observers have termed "patriarchal": a systematic pattern of age- and gender-

12. This chapter draws on the oral histories conducted in 1984 with Dagua residents, informal discussions and formal interviews conducted in El Palmar in 1978–79, Dagua parish records, and a number of secondary sources, including such important historical works as Ortega's (1920) two-volume history of railroad construction in Colombia, and Vergara y Velasco's three-volume, turn-of-the-century geographical history of Colombia (reprinted in 1974), as well as the research of faculty and graduate students at the Universidad del Valle, particularly Professors Germán Colmenares, José Escorcia, and Margarita Pacheco.

based inequalities in access to resources, to income, and to household decision-making power, with the male head of household as the ultimate authority. The fundamental conditions of existence of these family farms were access to land, internal relations that ensured a family labor supply, and technical and market conditions (both product and labor market) that enabled the household to obtain at least the minimum subsistence needs for intergenerational reproduction of the household unit.[13]

A number of important points emerge from this study of the development of agrarian structure in the mountains of nineteenth-century Colombia. First, I suggest that the limits to the spread of family farming were not as severe as the literature suggests. This study of the Dagua region indicates a rich new area of research in the reconstruction of the roots of family farming in the rest of Colombia's mountain zones.[14]

Second, the investigation sheds some light on the forces that generated this nineteenth-century migratory movement. The literature tends to emphasize "push" factors stemming from extensive land monopolies in the most densely populated regions of the country. The histories of some of the early settlers of El Palmar reveal that for some migrants the principal "push" factor was not land scarcity, but rather the desire to escape from their subordinate

13. It should be noted that the land and market conditions are long-run rather than short-run conditions, in the sense that a patriarchal unit just starting out might well begin with inadequate land for household subsistence. In fact, many of the settlers of the Dagua region began as agricultural wage laborers and gradually acquired land, so that a cross-section view at any point during this early settlement period would have revealed landless households whose subsistence derived wholly from wage labor, and farm households ranging from those combining wage labor with the cultivation of small plots to those with large holdings employing wage labor. It is necessary therefore to establish the overall or average conditions of existence for the family-farm form of organization, rather than to demonstrate that those conditions existed for each individual family unit.

14. There is increasing recognition of the scale of nineteenth-century colonization movements throughout Colombia (see, for example, Villegas 1977; Fajardo 1981; Parsons 1968; and López-Toro 1970 on Antioquia). With the exception of studies of Antioqueño communities of coffee growers, however, little is known of the fate of the thousands of settlements created during this great migratory movement. Because recorded information about smallholding agriculture exists primarily in documents related to conflicts between migrants and large landowners, recent efforts to reconstruct the changing agrarian structure of the nineteenth century have emphasized the growth of large landholdings at the expense of family farming. A comprehensive study from this perspective is LeGrand 1980.

positions in the patriarchal households of their birth. The analysis also provides some insight into the "pull" factors operating in colonization zones such as the Dagua region, where there was not only land but also an emerging structure of local and regional markets. Many of the major products marketed today, such as block brown sugar and coffee, were sold by the earliest settlers. These products and others entered a complex marketing system, whose evolution has closely paralleled the development of transportation routes through the region. These markets offered commercial opportunities for those who acquired land, while the market production spawned a wage-labor market that enabled poor migrants to accumulate funds and ultimately acquire land themselves. By establishing the historic economic significance of family farm market production in this region of the Andes, I refute the commonly held view that these early smallholding communities were static, subsistence economies insulated from the effects of market forces.

While the conditions of existence of the family farm were established in at least some regions of nineteenth-century Colombia (an important task now is to determine how extensive that process was), these conditions were far from permanently assured. The history of El Palmar, from its settlement in the late 1800s to the present, reflects a process of continual change and disequilibrium. This disequilibrium arose not only from changes in the general market environment to which the farmers of El Palmar responded (the "external" dynamic of the country's economic growth and development), but also from forces for change operating within the community itself (the "internal" dynamic of population growth and resulting fragmentation of holdings).

Subsequent chapters move back and forth between micro (household and community) and macro (regional, national, and international) levels of analysis in order to examine all the factors affecting the conditions of existence of Colombia's mountain smallholders over the course of the twentieth century.[15] In chapter 3, I consider the national economic transformation and mounting rural violence of the 1920s and 1930s and show that, despite these changes, family farming remained competitive with estate production in this pe-

15. These chapters draw on the 1978–79 El Palmar household interviews, Colombian population and agricultural census data, and a number of secondary sources.

riod. Indeed, estate production of many crops was itself based on the internalization through various tenure arrangements of the patriarchal household form of production, rather than on a unique form of large-scale production that could give it an advantage over family farming. In this chapter, I develop a theoretical argument for the competitiveness of the family farm based on technical characteristics of agricultural production and their compatibility with the labor process of the patriarchal household. Many branches of agricultural production are by nature biological, sequential, spatial, and seasonal processes. As a result, they are characterized by diseconomies rather than economies of scale, and they therefore give a technical advantage to small-scale, family-labor-based production, while posing obstacles to the development of large-scale capitalist production.

Despite this continued competitiveness of peasant production, internal processes were leading to growing difficulties for this sector, difficulties that I explore in chapter 4. Population growth, in the context of a closing mountain frontier by the 1920s and limited nonagricultural employment opportunities in the following decades, led to fragmentation of holdings, more intensive use of local resources, and mounting problems of soil erosion and exhaustion. It became increasingly difficult for the El Palmar households to obtain a livelihood from their agricultural production or local employment. By the 1960s migration became an important alternative. Similar processes were occurring in other family-farm communities around the country.

Furthermore, as chapter 5 shows, the Colombian estates were finally beginning to modernize in the 1950s, shedding their reliance on peasant tenant production and adopting capital-intensive methods of large-scale, wage-labor production. Modern coffee technology also began to be widely disseminated in the 1960s. Even in the face of these new changes, the family-farm sector continued to play an important role in the national economy in the production of domestic staples, since the modernizing estates were concentrating on the production of a variety of other products. Small-scale coffee production also remained significant. A contradiction was therefore developing between the continued national production role of the family farms and the increasing inability of this agricultural production to support the farm households.

Chapter 6 examines the emerging response to this contradiction in the community of El Palmar, and in the nation as a whole. The regional economic development of the 1970s, spurred by recovering coffee prices, provided a more favorable market environment for family-farm production in the Dagua mountains. In addition, the farmers had greater access during this period to technological changes that increased the productivity of their small plots. Off-farm income-earning opportunities also increased, providing the households with the ability to diversify their livelihood strategies. This chapter takes the study of El Palmar up to the 1983–84 agricultural year.[16]

The material in chapter 6 demonstrates how the conditions of existence of the family-farm sector (land, labor, markets, economic competitiveness) have been unevenly affected by economic development in Colombia over the past several decades. While some conditions have been strengthened (for example, the increasing substitution of purchased inputs for deteriorating land resources), others have been progressively eroded (for example, the weakening of internal household relations that assured a "cheap" family labor force). Despite this unevenness, I conclude, these family farms continue to be competitive producers of a number of agricultural commodities. I show how the farmers of El Palmar have selectively adopted new crops and technology, including the new coffee technology, and demonstrate the positive effect on net farm income of these changes, as well as the broader improvement in the economic status of El Palmar households in the 1970s. The economic changes of this period were beneficial to land-poor households as well as to peasant smallholders.

The experience of this community refutes the position that technological change and increased commercial integration will necessarily destroy the competitive position of peasant producers. In El Palmar, these developments increased peasant participation in agricultural product markets. Nevertheless, I show that absolute poverty is still a serious problem in the community, and demonstrate that this persistent poverty reflects the limitations of technological and commercial development in the context of the fundamental problem of peasant land scarcity. The conclusions drawn

16. Based on a household survey conducted in January 1984.

from El Palmar are generally reinforced by the results of Colombian government programs of the 1970s that attempted to increase peasant production and income through the introduction of technological innovations in smallholder communities. Although the changes of the 1970s increased peasant participation in agricultural product markets, rural householders are increasingly becoming part-time agriculturalists who rely on a variety of income sources. The rural contradiction therefore persists in Colombia.

Chapter 7 considers possible future trends in the conditions of existence of the family farm in El Palmar and in Colombia as a whole. A key issue in this analysis is the question of the basis for the continued economic competitiveness of family-farm production. This brings us back to the theoretical issues raised in chapter 1. The case of El Palmar suggests that the behavioral cost advantages postulated by the Chayanovians will not continue to be a basis for the persistence of family-farm production. Although these were significant in the past, they are gradually being eroded by the spread of market relations and the related weakening of patriarchal authority over the family labor force. I argue instead that the underlying basis for the continued competitiveness of family-farm production is to be found in the technical characteristics of agricultural production.

However, the technical cost advantage of family farming is not a sufficient condition for the existence of this form of production. Economically competitive units may be forcibly dispossessed, or the state may intervene to support an economically noncompetitive form of production. Furthermore, an increasingly important aspect of state policy vis-à-vis agrarian structure is its role in the development of agricultural technology, which can alter the "natural" conditions of agricultural production in favor of one or another organizational form.

After examining each of the underlying conditions, I conclude that the persistence of the economic competitiveness of the family farm in the future depends on economic, political, and ideological developments that are socially constructed, and therefore cannot be projected with certainty. While not providing an unequivocal prediction, the analysis does provide an understanding of the specific factors likely to influence the future course of peasant production in Colombia. The central project of this book is to redirect our analysis of peasant production out of the confines of the existing paradigms and situate it instead within a broader theoretical and methodological framework, one that emphasizes the analysis of the technical and social factors that condition agrarian structure in each concrete case. Only through such a comprehensive theory of agrarian development can we approach the study of the "peasant question."

The case study presented in this book thus serves a threefold purpose. First, it provides concrete historical information on family farming in one region of the Andes, providing an example of the important role that smallholding production has played in the economic life of the country. The study therefore contributes to the recent efforts to reconstruct Colombia's agrarian history. Second, the study of this one concrete instance enhances our theoretical understanding of the conditions of existence of family farming, and leads to a number of conclusions regarding the past, current, and future status of family farming in Colombia. Finally, the study illustrates the usefulness of a nondeterminist approach to the analysis of the "peasant question," and it therefore has theoretical and methodological implications beyond the study of a particular region or a particular country. I hope that other researchers interested in the analysis of agrarian structure, whatever their geographic focus, will find this approach useful.

*Chapter One*

# The Peasant Question: Theoretical Perspectives on Peasant Production and Their Application to the Latin American Peasantry

The social and political significance of the peasantry in Latin America has long been recognized.[1] Its economic importance, however, tended to be overlooked in postwar development models that assumed the economic superiority of large-scale production and the labor-absorptive capacity of industry, backed up by a conception of peasants as subsistence producers with few market links.[2] By the early 1970s this perspective had altered considerably: peasant production was taking center stage in official and academic discussions of Latin American agriculture.

The new emphasis on the economic importance of the peasantry was due in large part to the increasing difficulties of the agricultural sector. Latin America was not the only region of the Third World to have encountered such difficulties. Per capita agricultural output was stagnating overall in the developing countries. The rate of

1. This has been particularly true since the important role of the peasantry in the Mexican Revolution (which has generated a substantial body of literature) and the growth of peasant movements in many Latin American countries in the 1920s and 1930s. See, for example, Stavenhagen 1970.

2. An excellent example of this perspective is the report of the influential 1949 World Bank mission to Colombia (Currie 1950).

growth was only .6 percent between 1948–52 and 1960, and there had been zero growth overall in the 1960s (Todaro 1985:288, table 10.2). At the same time, evaluations of the Green Revolution programs—programs launched with great hope in the 1960s as the answer to Third World rural poverty and hunger—pointed by the early 1970s to their shortcomings and weaknesses (Falcon 1970; Cleaver 1972; Griffin 1974).

Beneath the apparent successes of the 1960s in the initiation of a process of commercial agricultural modernization in many Third World countries lay a submerged, but growing, crisis. It began as a crisis of rural poverty and hunger, as the growing rural population crowded onto an increasingly marginal and deteriorating land base. By the 1960s and early 1970s, it had become a crisis of rising urban food prices and of growing food imports, as Third World cities bulged with the massive exodus from rural areas. These structural weaknesses were brought into sharp focus by the events of the early 1970s. The oil price increases, with their impact on the price of petroleum-based agricultural inputs and on world food prices, and the concomitant drought and famine in the nations of the African Sahel dramatized to the world the vulnerability of Third World agriculture.

The solution to these difficulties was focused increasingly on Third World peasant producers, in what might well be called the "decade of the peasantry." The World Bank formalized a pro-peasant emphasis in its lending policies with Robert McNamara's 1973 Nairobi speech (McNamara 1973). International donors and Third World governments endorsed rural development projects aimed at peasant producers. As these programs proliferated, the academic literature and that of international development organizations began to analyze these programs on two important levels: the economic rationality of targeting small-scale producers, and the impact of the programs on those producers. In other words, these programs began to be analyzed both theoretically (is there an economic argument for channeling scarce resources to the development of peasant food production rather than large-scale production?) and in practice (did these programs in fact increase peasant marketed output and improve producer incomes, and did this contribute to overall decreases in rural poverty?).[3]

3. A comprehensive overview of these programs and their evaluations is provided by Crener et al. 1984. The fact that these analyses were developed after the

These discussions formed part of a larger debate on the basic status of peasant producers today. Do peasants continue to play a significant role as agricultural producers, or has their competitiveness been destroyed by recent changes in the economies of Third World countries? If they can no longer compete, then how do we account for the large numbers of smallholdings that remain in those countries? In Latin America, for example, holdings of less than five hectares (twelve acres) accounted for over half of all agricultural holdings in a number of countries during the 1960s and early 1970s, including Brazil, Colombia, Mexico, and Peru (de Janvry 1981:122). A United Nations Economic Commission for Latin America (CEPAL) study found that almost four-fifths of all agricultural units in Latin America in 1980 were "small-holdings" or "family units" (López Cordovez 1982:26). What are these small farms like, and how can their persistence be explained?

## Theories of Peasant Production and the Development of Capitalism

The debate over the impact of economic development on the peasantry, the "peasant question," has brought forth a number of contending theoretical views. Two perspectives have been particularly prominent in recent discussions of the Latin American peasantry. One view holds that there are aspects of the decision-making behavior of peasant producers that enable them to compete with large modern farms. It is this economic competitiveness of peasant producers that explains the persistence of small farms in Latin America. This view is called the *campesinista* (peasant persistence) position. According to this position, peasant agriculture is "stable": able to retain its basic characteristics over time despite broad changes in the economic environment. The second view, the *descampesinista* (peasant elimination) position, is that large farms are more efficient than the peasant units and are coming to dominate agricultural markets in Latin America. The peasants are being forced increasingly to rely on wage labor for their livelihood. The

initiation of the programs rather than before underscores the fact that the pro-peasant policies were undertaken without a prior theoretical basis and with little understanding of the concrete conditions of the target population.

small peasant holdings are not really farms, but rather food plots used to supplement wage income.[4]

## *Arguments for the Persistence of Peasant Production*

Writers from a wide variety of disciplinary and political perspectives have argued that peasant production can persist (and has persisted) under the processes of broad societal transformation referred to variously as "economic development" or "capitalist development."[5] Their common definition of "peasant production" is the use of household labor in agricultural activities for the purpose of providing all or part of household livelihood. What is common to their vision of the economic development process is the growing supremacy of market relations in the economy as a whole, so that all products tend to become "commodities" sold on the market, and all resources also exchange freely on the market. In this development process, the peasant producer is gradually drawn into market production. Thus the writers whose work I review here all analyze peasant households whose agricultural activities are well embedded in a developed market context, as opposed to non-market-oriented or "subsistence" peasant production. These writers all argue that peasant commodity production in a developing market economy is consistent with the stability and persistence of peasant units of production over time.

Despite these commonalities, there are important differences among members of the "peasant persistence" school. One difference is in terminology, with some using the term *peasant* and

4. See Heynig 1982 for a review of these positions that follows along the same general lines of categorization and analysis as the review in this chapter.

5. I focus here on those works that have particular relevance to the debate over the "fate" of peasant producers, rather than presenting an exhaustive review of the voluminous literature on the peasantry. The works reviewed deal with various aspects of the "supply" side of peasant agriculture—the nature of the peasant farm as a production unit and the nature of the distribution of peasant products. I leave aside another body of works that emphasize the "demand" side of the peasant unit—the nature of the demand for consumer and producer goods emanating from the sector of peasant producers—which, while it raises important issues, is not directly relevant to the questions under consideration here. Johnston and Kilby (1975), for example, have emphasized the positive implications for economic development of the broadly based demand structure that they argue is a consequence of a family-farm agrarian structure.

others using *family-labor farm* or simply *family farm*. Others, most notably Friedmann (1978, 1980), have attempted to distinguish between subsistence household production, designated as "peasant," and market-oriented household production, referred to as "simple commodity" production.[6] While these distinctions are frequently semantic, they do reflect differences in analytic emphasis, as I shall show. These writers also come to widely different conclusions regarding the effect of the developing market economy on the farm household's standard of living and on the technological development of peasant production.

*The Anthropological Perspective.* One branch of the persistence-of-peasant-production approach is rooted in traditional social anthropology, which sought to develop a methodology for the analysis of the social characteristics of isolated tribal groups. When anthropologists turned to analyses of the peasantry, the emphasis on self-contained cultural and economic entities gradually gave way to a recognition of the many links between the peasant communities and the broader economy and society (Foster 1967), leading to such characterizations of the peasantry as Kroeber's (1948) "part society and part culture." Within this context of external relations, however, the anthropological focus on the distinctive features of peasant society has remained the conceptual linchpin of an approach to peasant studies that seeks to define a specifically "peasant" economy.

This approach is well represented by the work of Teodor Shanin, who refers to "four essential and interlinked facets" of peasant economy: "the family farm as the basic multi-functional unit of social organisation, land husbandry and usually animal rearing as the main means of livelihood, a specific traditional culture closely linked with the way of life of small rural communities, and multi-directional subjection to powerful outsiders" (Shanin 1974:63–64). He argues not only for the specificity of peasant economy, but also for its resistance to outside pressures and its ability to persist in widely varying socioeconomic contexts.[7]

6. Simple commodity production is a category in political economy that refers to small-scale, household-based production for the market, as distinguished from "capitalist" production, which is also market-oriented but is based on the use of wage labor. One important area of controversy, to be considered below, is whether the absence or presence of the profit motive also distinguishes these two forms of production.

7. "While the impact of the broader societal organisations . . . 'encircling'

In his analysis, Shanin (1974:67) postulates that peasant production, based primarily on the use of family labor, is geared to family consumption needs and the need to meet outside claims on the peasant household (rents, taxes, and other exactions). While peasant households might want to earn more than these basic needs, Shanin argues, they are unable to do so because of "leveling mechanisms" that stem from factors both internal and external to the peasant economy. These mechanisms include partitioning of holdings, random effects of nature, migration, egalitarian norms in some cases, and exactions by outside groups, all of which continue to operate today on the peasant economy (Shanin 1974:193). As a result of these homeostatic forces, the more or less equal distribution of resources between peasant households is maintained over time, and the peasant economy is therefore not only specific in its behavioral characteristics, but also stable (Shanin 1974:71). This explains the continued existence of peasant farms. However, these farms will tend to be technologically stagnant and poor, since the leveling mechanisms prevent capital formation in the peasant economy.[8]

*The "Chayanovian" Perspective.* Many writers who argue for the persistence of peasant production draw on the work of A. V. Chayanov, an agricultural economist whose research group studied conditions in the Russian countryside in the early 1900s.[9] From these observations, Chayanov developed a theory of peasant decision-making behavior. He argued that the goal of peasant production is to provide a livelihood for the household. The "profit-motive" is absent from the peasant farmer's calculations, whether

---

peasant family farms has been unmistakeable and powerful, yet it did not destroy certain 'generic' major similarities of peasant economy and social structure in different parts of the world. Peasant economies have indeed shown a remarkable degree of structural persistence under different external impacts, their essential characteristics outliving, as it were, most of the social and economic systems in which they appeared" (Shanin 1974:186).

8. Eric Wolf (1966) provides a similar analysis of peasant economy and its subjection to external leveling mechanisms.

9. Two of Chayanov's works were translated into English in the 1960s and have been very influential in attempts to develop a theory of peasant production: "Peasant Farm Organization" and "On the Theory of Non-Capitalist Economic Systems," in Thorner, Kerblay, and Smith 1966.

the household produces only for its own consumption or produces some output for the market. The level of production is determined not by its profitability, but by the consumption needs of the household and by the drudgery involved in the necessary farm labor. If prices for its products fall, the family may actually work more in an effort to maintain its level of consumption. From these basic characteristics, Chayanov developed two lines of analysis: to show that the peasant economy was internally stable and that it was competitive with other types of agricultural units.

In Chayanov's analysis, the peasant farm depended primarily on family labor (to emphasize this fact, he called these units "family-labor" rather than "peasant" farms). Any expansion of output therefore meant that family members would have to work harder. The household, in deciding on its scale of operation, would compare the (diminishing) marginal utility of additional consumption to the (increasing) drudgery associated with increased production. Although the scale of operation for a household at any particular point could be affected by a number of factors, Chayanov felt that the key factor was the ratio of household members ("consumers") to its working-age population ("producers"). As this consumer/producer ratio changed over the household's life cycle, the sown area of the farm would be expanded or contracted. According to Chayanov's model, the drudgery of additional labor in an essentially family-labor enterprise would prevent wide differences from emerging between households in the levels of production and consumption per capita, once variations in the consumer/producer ratio had been taken into account.

Chayanov thus developed a model of self-replicating family-labor farms, geared to household consumption and limited in their production goals by the drudgery of labor. As with Shanin's peasant economy, Chayanov's economy of family-labor farms was stable over time.[10] In Chayanov's model, however, stability was due essentially

10. Chayanov presented evidence from the Russian agricultural statistics to argue that his theoretical family-labor farm was in fact an appropriate model with which to characterize the Russian peasant household. He argued from cross-sectional statistics that differences in sown area and in production per family member in the Russian countryside were correlated with differences in the consumer/worker ratios of the households. As a result, differences between households did not represent a process of "social" differentiation (a cumulative development of systematic wealth and income differences between peasant households),

to internal behavioral characteristics, rather than to outside factors that "leveled" differences between households.[11]

Nevertheless, Chayanov was explicitly concerned with one fundamental aspect of the relationship between peasant production units and the broader economy: the issue of the competitiveness of peasant production with "capitalist" farms. The capitalist farm is a business, not a way of life. Its owners invest in it because they hope to earn a rate of profit at least as great as the rate they could earn elsewhere. The level of production is determined by its profitability. If profitable, that money will be reinvested and the farm will expand. If not, the owners will eventually sell off the property and invest elsewhere. The capitalist farm will be completely market-oriented, use hired labor (which it can obtain only by paying the going or market wage), and tend to be larger than the family-labor farm.

Chayanov argued that the family-labor farm, whose owners were willing to forgo profits, could accept lower product prices than the capitalist farm, whose owners would shut down their operations if they could not earn a profit. If necessary, the peasant household would even "super-exploit" itself and accept a return per hour of labor that was lower than the market wage, something the owners of a capitalist farm could not do. In situations where land shortages prevented family-farm households from expanding sown area to meet increasing consumption needs, they could maintain consumption levels by switching to more labor-intensive crop mixes—another form of increasing annual labor returns while decreasing

---

but rather a cyclical and stable process of "demographic" differentiation related to the life cycle of the peasant household. Chayanov supported this argument by presenting longitudinal data to demonstrate a cyclical process of small-farm growth and large-farm (in the peasant context) breakup in the Russian countryside between 1882 and 1911 (1966:67).

11. A number of social scientists have based their analyses of present-day peasant communities on a similar model of the family-labor farm, using a cross-sectional approach to show a correlation between household composition and such economic variables as sown area or income per capita. In these analyses, the persistence of peasant farms is explained in terms of their presumed behavioral characteristics (the consumption and drudgery limitations on farm expansion), and demographic differentiation is demonstrated. See, for example, the collection of "Chayanovian" studies in Durrenburger 1984. In general, these studies emphasize internal homeostatic forces, leaving aside questions of the broader economy and society with which the peasant community is integrated.

returns per unit of labor (Chayanov 1966: 113–15).[12] The family-labor farm could therefore compete well with the capitalist farms, sometimes even driving them out of business. I shall refer to this argument as the "behavioral advantage" theory of the persistence of peasant production.

Chayanov (1966: 257–63) concluded that the development of agricultural markets would not lead to the displacement of peasant farmers by large modern farms, but instead to the incorporation of peasant producers into a complex system of commercial agricultural production and distribution. Peasant producers would become highly dependent on producers of agricultural inputs, banks, and agricultural processing industries, all of whom could force the small farmer to agree to disadvantageous terms. Chayanov called this the "vertical concentration" of agriculture under capitalism, and he argued that peasant farmers would need to form cooperatives to give themselves greater bargaining power in the marketplace.

*The Marxist Perspective.* Chayanov's analysis of the economic position of the family farm under capitalism has been emphasized by a number of Marxists writing on the persistence of peasant production in France, Greece, Africa, and elsewhere (Servolin 1972; Amin 1974; Amin and Vergopoulos 1974; Vergopoulos 1975, 1978; Bernstein 1977, 1979). Bernstein develops the "vertical concentration" argument in the case of African peasant producers, detailing the numerous ways in which they have been integrated into commercial trading networks. This includes trade organized by merchants, industrial users of peasant products, industrial producers of agricultural inputs, financial institutions, and the state. Bernstein argues (1977: 68–72) that these processes are all forms of increasing control over the "conditions of production" of the peasantry by various forms of capital.

As to why peasant production should be thus conserved rather than eliminated by competition from agricultural production on capitalist units, Bernstein (1977: 62–63, 66, 72) argues along Chayanovian lines from the consumption goal of peasant production to

12. Chayanov argued that this was the case in a number of European countries (1966: 115–16).

the ability of peasant producers to compete with capitalist production by accepting a return below the market wage (the "devalorization" of labor time). He also argues that the continued production by the peasant household of agricultural output for direct home consumption ("use-value" production) can allow the family to reduce the level of remuneration they will accept on commodity sales even further.

In an important essay, Amin (1974) has expanded on the general Chayanovian model of the behavioral competitiveness of the family farm with a discussion of the issue of land rent.[13] Amin follows Marx in distinguishing between "differential rent" (differences in return per acre due to differences in soil quality, location with respect to roads and markets, etc.) and "absolute rent." In Amin's analysis, "absolute rent" is a portion of the agricultural surplus that accrues to landowners because of the existence of private property in land. It would accrue to owners of even the most marginal piece of land, land whose differential rent would be zero. As such, absolute rent is a subtraction from the surplus that would otherwise accrue to capitalist agricultural producers (a subtraction in the form either of direct payment of rent or of an elevated purchase price of land).

Amin argues that there is no objective basis for the determination of the level of absolute rent. Rather, it is determined subjectively, through the resolution of the political struggle between landowners and the emerging class of capitalist producers. This depends in part on the character of the landowning group. The existence of a group of individuals whose income derives from control over land rather than productive activity—that is, the remnants of a "feudal" landlord class—implies the continued existence of absolute rent during capitalist development unless their political power can be weakened.[14]

The peasant producer, on the other hand, does not relate to land in the same way as the landlord. For the peasant, land is an "instrument of labour" (Amin 1974:7) rather than a source of rent. Amin

13. This essay was published in a revised form in Amin and Vergopoulos 1974.

14. This could occur through a capitalist revolution, as described by Barrington Moore (1969) for a number of industrial countries, including the United States and England. Unless the political power of the landlord class can be eroded, absolute rent will persist even while the development of agricultural technology is eroding the basis of differential rent.

(1974:27) summarizes Chayanov's analysis of the behavioral logic of the family farm and argues, with him, that the goal of peasant production is "to live off the land." The peasant will therefore forgo absolute rent as well as profit.

This analysis leads to two conclusions regarding the political forces that may support the preservation of peasant production. One is that the existence of a politically powerful landed elite is an obstacle both to the development of capitalism in agriculture (since productivity would have to be great enough to cover absolute rent as well as costs of production) and to the overall development of capitalism (since it raises agricultural prices and therefore lowers profits in all spheres of capitalist production). This is the basic position of the structuralist analysis of agrarian stagnation in Latin America and its implications for industrial development.[15] The second conclusion is that an agrarian structure of family farmers is preferable from the perspective of the overall development of capitalism. It is not surprising, then, to find "modernizing" groups supporting land reform, as in the case of the structuralists in Latin America. By abolishing the class of property owners that demands land rent, and replacing it with a group of property owners who do not, the prices of agricultural products can be lowered.

Amin concurs with Chayanov's analysis of the competitiveness of the family farm vis-à-vis the capitalist farm (due to the ability of the family farmer to forgo profit) and the development of forms of vertical concentration that are "mechanisms of domination by the capitalist mode over the peasant economy" (Amin 1974:31). With his analysis of land rent, however, Amin goes further than Chayanov to argue that the family-farm structure enables "capital [to] derive not only a better overall rate of profit but also a better political control over society" (1974:34).

Amin differs sharply from Chayanov in his analysis of the implications of this process for the standard of living of the farm household. He draws a strong conclusion regarding the process of vertical concentration: not only is the family farmer *willing* to forgo rent and profit and accept compensation only equal to the agricultural wage, but the process of vertical concentration *forces* prices

15. See, for example, the brief summary of structuralism in de Janvry 1981:146.

of peasant products down to this level. As a result, the family farmer is "proletarized": "on the surface he remained a commodity producer who offered his products on the market, but in actual fact, he was a seller of labour power. . . . Thus the peasant was actually reduced to the status of a person working at home under the domestic system" (1974:30). Amin (1974:44) refers to such producers as "semi-proletarian." In extreme cases, furthermore, the level of remuneration can be driven below the "value of labour-power," that is, below the level of minimum subsistence for the household. Amin (1974:43–47) calls this the "over-exploitation of the peasantry," and cites peasant commodity production in sub-Saharan Africa as an example, with its toll in terms of soil exhaustion, malnutrition, mortality, and famine.

In this essay, Amin draws together the perspectives of Chayanov and Shanin. He builds on Chayanov's concepts of family-farm organization, which imply its economic competitiveness with other forms of agricultural production and its persistence. At the same time, he expands the external "leveling mechanisms" of Shanin's analysis to include vertical concentration as the form of external exaction on the peasantry under capitalism, in a manner similar to Bernstein's discussion. As a result of this mechanism (whose exact nature is not clearly specified), Amin concludes that the income of the peasant producer is kept down to the level of the wage worker. Peasant production is thus both competitive—and therefore "functional" to the capitalist sector as a means of assuring "cheap food" (Amin 1974:39)—and stable.

These theoretical arguments have been expanded by Amin and Vergopoulos (1974) in a two-essay volume, and applied by Vergopoulos (1975) to the study of agrarian structure in Greece.[16] Vergopoulos examines the precise nature of the external exactions and leveling mechanisms operating on Greek peasant producers. Unlike most writers of the peasant school, however, he argues that the reduction of peasant income to the level of wage income that results from these mechanisms does not preclude investment on the peasant farms. Rather, these pressures may force family farmers to adopt productivity-increasing techniques in their efforts to

16. The summary of Vergopoulos' (1975) arguments presented here is drawn from Mouzelis 1976. See also Vergopoulos' (1978) restatement of his thesis in response to critiques by Mouzelis and others.

maintain household consumption. These investments raise productivity, but not net income, because of the continued operation of the leveling mechanisms.

*The Neoclassical Perspective.* Neoclassical economists, too, have argued that the family farm can be compatible with processes of economic development. Some, concurring with the Latin American structuralists, have suggested that the family farm may even be preferable to large-scale production (Johnston and Kilby 1975; Dorner and Kanel 1971; Owen 1966). Owen, in a manner similar to Vergopoulos', bases his argument on the process of technological change on the family farm. He argues that the "Mill-Marshallian" treadmill forces the family farmer to adopt new techniques, but also transfers the resulting productivity gains away from the family farm to the rest of society through the agriculture/industry terms-of-trade (Owen 1966:5). The competitive nature of the family-farm sector and the inelasticity of demand for food are principal factors in this transfer process, operating as different "leveling mechanisms" from those reviewed above.

The argument with respect to the receptivity of the family farmer to technological change has been bolstered by studies demonstrating that peasant producers allocate resources in an economically "rational" manner. These include studies carried out by neoclassical economists such as Schultz (1964, 1965) and Hopper (1955) and by economic anthropologists such as Tax (1953) and Nash (1966). Their concept of peasants as profit maximizers is in marked contrast to the Chayanovian vision of producers whose output goals are limited by the decreasing utility of consumption compared to its drudgery costs. The introduction of profit maximization, in the context of technological change and the use of hired labor, opens the door to unlimited growth of the production unit and thus destabilization of the peasant economy.[17]

The neoclassical pro-peasant position sidesteps this fundamental

17. This is also possible under the Chayanovian behavioral model, given the presumption that marginal utility of consumption is always positive and given the possibilities for drudgery reduction through certain types of technological change and through the use of hired labor. Chayanov sidestepped this issue through his emphasis on family-*labor* farms, where drudgery is always borne by the family. Other Chayanovian-based analyses emphasize external mechanisms that prevent expansion.

issue of the stability of the peasant economy by emphasizing the empirical evidence for an inverse relationship between yield and farm size in Third World agriculture. The implication is that small-scale production units are economically competitive with large units. The argument is taken to the point of advocating land reform based on the observed higher yield of smallholdings (Dorner and Kanel 1971; Berry and Cline 1979). However, the issues of the competitiveness and stability of peasant production are not explored on a theoretical level, as they are in the Chayanovian-type models of specific peasant behavioral logic.[18]

*Contrasts in the Postulates of Peasant Persistence.* We can summarize these various arguments for the persistence of peasant production under the development of capitalism in terms of two principal—sometimes contradictory, sometimes convergent—lines of analysis. One approach is grounded in the postulate of the specific behavioral logic of the peasant farm. From this is derived the *competitiveness* of peasant production vis-à-vis large-scale capitalist agricultural production. In some variants of this approach, the *stability* of the peasant economy is also derived from these behavioral postulates. In other variants, stability is derived not only from the limited production goals of the family farm, but also from the operation of external pressures, which act as leveling mechanisms that prevent unlimited growth of individual family-farm units. In some variants, technological change and productivity increases are considered to be incompatible with the peasant farm because of the extent of external pressures; in others, technological change is considered to be possible and perfectly compatible with the consumption motive of the peasant farm.

The second pro-peasant approach is grounded in the postulate of peasant profit-maximizing behavior—that is, it argues against the specificity of peasant production behavior. Nevertheless, the peasant farm is argued to be competitive based on empirical evidence of higher yields on small farms. Because of its profit motive, peas-

18. By arguing that the observed outcome of peasant production decisions is consistent with neoclassical profit-maximizing behavior, the question of whether peasants are *in fact* maximizing profit is avoided—i.e., one need not consider whether peasant producers impute labor costs and whether they are ultimately limited in their production goals by utility and drudgery considerations (once their income position moves beyond basic subsistence).

ant production is considered to be compatible with technological change that offers possibilities for increased profit.[19] This approach does not consider the question of the stability of the family farm or the theoretical basis for its observed competitiveness.

## *The Case for the Disappearance of Peasant Producers*

At the same time that Chayanov was developing his theory of the persistence of peasant production under the impact of economic development, other analysts were drawing diametrically opposite conclusions. The "classic Marxist" argument as developed by Kautsky and Lenin was that capitalist farms would come to dominate the agricultural sector (Marx 1857; Kautsky 1902; Lenin 1920). The capitalist farms would be able to produce at lower costs per unit than the peasant farms, both because the capitalist farms would be much larger and because they would adopt modern technology. The prices that would result in agricultural markets dominated by capitalist producers would be so low that peasant producers would be reduced to a miserable standard of living.

Most would turn instead to wage labor on the capitalist farms, selling or losing their land and becoming "proletarians." A few peasant producers, those who had more land to begin with, might be able to take advantage of the new technology and the difficulties of their poorer neighbors to become capitalist farmers themselves. Their farms would become large, modern, wage-labor-using businesses. Lenin called this the "farmer road" to the development of capitalism in agriculture. In other cases, large traditional or "feudal" estates would be the locus of capitalist development as their owners modernized production (the "Junker road"), or urban merchants or professionals might buy or rent agricultural property to initiate modern farming operations (the "merchant road").

A number of Marxist scholars have applied this classic analysis to today's debate over the fate of the peasantry. Patnaik's (1974) critique of both Chayanovian and neoclassical "subjectivist" positions, which she groups together under the category of "Neo-Populist," is representative of this school. Patnaik bases her critique on an

19. Some analyses consider not only expected profitability, but also the riskiness of the new techniques—that is, the variance around the expected outcome.

analysis of Chayanov's theoretical arguments for the competitive position of the family-labor farm vis-à-vis capitalist units of large-scale agricultural production.[20]

Chayanov provided a numerical example to illustrate his argument. He gave hypothetical costs of production on a capitalist and a family-labor farm, and then considered the competitive position of the two farms at different price levels. At a high price level, the capitalist farm would be earning profits and the family-labor farm would be enjoying a return per unit of labor greater than the wage on the capitalist farm. At a lower price, the profits of the capitalist farm would be eliminated, while the peasant farm household would simply be accepting a lower return to household labor (Chayanov 1966:88).

Patnaik points out that Chayanov's conclusions rest on his assumption that the family farm and the capitalist farm have the same production function (the same nonlabor costs, the same labor input, and the same physical output). She notes that "*all* the main theoretical propositions of the school: the superior viability of the family farm compared to the capitalist farm in the face of adverse market conditions, the superior 'efficiency' of the family farm in producing a higher output per acre, the non-operation of the law of capitalist concentration in agriculture because small-scale production is no less efficient compared to large-scale production—all

20. Although this is the key part of her critique, much of Patnaik's discussion is actually focused on a theoretical and empirical refutation of Chayanov's interpretation of the situation of the Russian peasantry at the turn of the century. She argues that the Russian agrarian statistics were selectively analyzed and presented by Chayanov to support his view that the Russian peasantry was a fairly stable group with little evidence of systematic differences in income and wealth. Patnaik argues, with Lenin, that in fact the agrarian statistics revealed a clear process of such systematic social differentiation (Patnaik 1979:379–86; Lenin 1920). Her critique in this section clearly reveals the complexities of the analysis of agrarian structure and agrarian change trends. In particular, her critique of Chayanov's statistical analysis of the relationship between household demographic and economic factors correctly focuses on the pitfalls of inferring causation from correlation (Patnaik 1979:388). This critique can be leveled as well, however, at the Leninist differentiation position: the cross-sectional analysis of rural statistics does not capture the underlying dynamic processes of mobility of individual farm units that lies at the heart of the Chayanovian homeostatic model. This point will be taken up again in the analysis of the El Palmar statistics. As a result, Patnaik's discussion of the Russian agrarian statistics is no more conclusive than was the original Lenin-Chayanov debate.

these propositions follow directly from the assumption of identical production conditions for family farms and capitalist farms" (Patnaik 1979:416). This, Patnaik argues, is a "fundamental fallacy." The emergence of the capitalist farm historically (according to Patnaik, "necessarily") takes place in the context of the development of agricultural technology that raises the productivity of labor in large-scale production. As a result, the capitalist unit may enjoy a productivity advantage equal to the capacity for self-exploitation of the family farm, allowing the two forms to coexist in the same market (Patnaik 1979:396–402).

This critique is applicable also to much of the recent empirical evidence presented by neoclassical advocates of land reform. In Latin America, the comparison between small-scale and large-scale agricultural production is based on data that do not distinguish between traditional estate production (the extensive operations of the latifundio) and modern capitalist agricultural units. While small-scale production may be competitive with the traditional estates, it is not clear from this data that it is also competitive with large-scale capitalist production.

Patnaik asserts that the "law of capitalist concentration" does operate in agriculture, once agricultural technology has developed sufficiently for capitalist agriculture to overcome the barrier of absolute rent. However, expansion of the *physical* scale of capitalist agricultural operations is limited by private property in land, giving rise to intensive concentration of capitalist agricultural production ("capital intensification") (Patnaik 1979:408–15). Specifically, she presents Kautsky's argument that the small farms may pose an obstacle to the physical expansion of capitalist agriculture because "the small family farms 'competed' with the capitalist farms using superior techniques, by working unbelievably hard for a return to labor which was below wages and by accepting a total consumption per capita which was below that of the farm worker" (Patnaik 1979:409).

Although not specifically argued by Patnaik, the logical outcome of the process of productivity-increasing investment in capitalist agriculture she envisions must be the complete displacement of peasant commodity production through the dual process of declining output prices and increasing wage-labor opportunities on the capitalist farms. Ultimately, the capitalist unit would be able to increase productivity to the point where it could cover its costs (including rent and normal profit) at a product price below the minimum acceptable level for the family farm, while wage labor would become an increasingly attractive alternative for household members.

Given the "inescapable" character of this process, is it possible to envision the persistence of any production on small-scale farms? Lenin and Kautsky proposed three possibilities (Patnaik 1979:409). One was that peasant production could persist because of the inadequate development of the labor market. Second, peasants might perceive nonpecuniary benefits to the less remunerative choice of farming over wage labor. The third possibility was that the inexorable development of the superiority of capitalist production would indeed destroy the commodity production role of the peasant farms and convert the peasantry into wage laborers—but into laborers who retained plots of land on which they produced for household consumption.[21]

The pressures of capitalist development may be resisted, then, by some "irrational" peasants who cling to their plots of land by retreating from commodity production to the sphere of use-value production for home consumption, despite the miserably low standard of living that this implies (Patnaik 1979:420n.34). The classic Marxist position envisions the ultimate conversion of the majority of the peasants into wage workers (while, depending on the form that the development of capitalist agriculture takes, a small number of peasant producers may become capitalist agricultural producers): "The old peasantry is not only 'differentiating,' it is being completely dissolved, it is ceasing to exist, it is being ousted by absolutely new types of rural inhabitants— . . . the rural bourgeoisie (chiefly petty bourgeoisie) and the rural proletariat" (Lenin 1964:177).

This argument is not the same as the "peasant persistence" argument that vertical concentration of agriculture under capitalism reduces peasant producers to wage-labor *equivalents* (i.e., reduces

21. Lenin (1920:180) argued: "Our literature frequently contains too stereotyped an understanding of the theoretical proposition that capitalism requires the free, landless worker. . . . The allotment of land to the rural worker is very often to the interests of the rural employers themselves, and that is why the allotment-holding rural worker is a type to be found in all capitalist countries."

their income from farm production to the level of agricultural wage workers). In this perspective, peasants retain a function as economically competitive producers of agricultural commodities. Competition between peasant producers and the pressures of vertical concentration force the prices of agricultural commodities down to the lowest level acceptable to the peasant household. If nonfarm income-earning opportunities are limited, this level will be the minimum level of subsistence for the household. If more remunerative nonfarm opportunities and/or insufficient farm income draw some peasants out of farming, the prices of agricultural commodities will rise, all else being equal. As the labor market develops, the long-term level of remuneration for peasant households should be the level of income of agricultural wage workers (the opportunity cost of the peasant's labor), although an income gap may persist because of peasant perceptions of nonpecuniary benefits of farming.

In the classic Marxist analysis presented by Patnaik, on the other hand, the peasants inevitably lose their competitiveness as commodity producers. The prices of agricultural commodities are set by costs of production on the more efficient capitalist farms. These costs eventually become so low that agricultural prices cannot generate an acceptable income on the family farms. As the labor market develops, peasants are drawn out of direct agricultural production to become laborers on units of agricultural production organized by capitalists, although they may retain access to small plots of land on which they produce household use-values. The decline in peasant commodity production has no effect on the prices of agricultural commodities, since these are set not by competition on the peasant farms, but rather by production costs on the capitalist farms. The conversion of peasant producers into actual wage laborers is not limited, as it is in the Chayanovian analysis, by an income-equalizing rise in the prices of agricultural commodities. *All* peasants are eventually converted into actual wage laborers, except those few who continue to resist at any cost (and survive by retreating completely from commodity production).

In the classic Marxist analysis, then, the extreme poverty of peasant producers derives from their attempts to compete with capitalist agriculture, rather than from the surplus drain that results from vertical concentration of agriculture. The policy implications of these two analyses are diametrically opposed. The classic Marxist analysis supports a policy of improving the position of peasants *as workers* in the economically preferable large-scale agricultural operations. The "neo-populist" analysis is frequently used to support a policy of improving the position of peasants *as commodity producers* through changes in marketing systems and channels of access to inputs (development of peasant unions, cooperatives, etc.).[22]

The argument that capitalist agriculture is changing peasants from commodity producers to use-value producing wage laborers, or "semiproletarians" (in a different sense from that used by Amin), has been developed most specifically in the case of Latin America by de Janvry and Garramon (1977), Deere and de Janvry (1979), and de Janvry (1981). De Janvry (1981) argues for the logical necessity of this emerging agrarian structure (capitalist agricultural production, peasant semiproletarian households) on two levels.

On one level, de Janvry criticizes the theoretical basis of the stability arguments of the pro-peasant position, the postulate that the expansion of the peasant production unit is inherently limited by consumption/drudgery considerations. According to de Janvry (1981 : 104), this cannot be "proved" by pointing to the actual low income level of the peasantry, since that income reflects the "extraction" of the "peasant's surplus labor . . . through a variety of channels." Despite their poverty, peasants may nevertheless be profit seekers.[23] De Janvry concludes that peasants are not a separate (noncapitalist, nonlabor) class, but are differentiating into those two classes. This is because

> the extractive mechanisms to which the peasant mode is subjected are market, and not individually, imposed and thus apply with equal intensity

22. The Marxist variant of the "persistence of the peasantry" position, however, argues from the impoverishment of the peasantry under vertical concentration to the possibility of a worker-peasant alliance for the revolutionary transformation of capitalism (since peasant producers are simply disguised wage workers under capitalism).

23. "Advocates of the peasant mode incorrectly use simple reproduction—an observed fact—as a defining behavioral characteristic of peasants because they confuse the inability of peasants to capture profits with their presumed nondesire for profits" (de Janvry 1981 : 104). In a similar manner, however, and in criticism of de Janvry's argument, the existence of "surplus extraction" cannot be seen to "prove" that, in the absence of such extraction, peasants *would* be profit seekers.

to all peasants. As a result, even if the community's surplus were fully extracted, positive surpluses would still be appropriated by the more favored and successful peasants while others would be driven below simple reproduction; and differentiation would occur (de Janvry 1981 : 103).

This conclusion is based on the rejection of the consumption/drudgery postulate (a rejection of Chayanov's internal leveling mechanisms), and its replacement by the postulate of peasant producers as profit seekers. De Janvry thus concurs with the neoclassical conception of the peasantry as rational, profit-maximizing producers. As such, they are capable of embarking on a process of unlimited expansion of the scale of their operation (assuming the existence of economies of scale, or at least the absence of diseconomies), an expansion that, de Janvry argues, is not prevented by generally unfavorable terms of trade or other extractive mechanisms operating on the peasantry as a whole (a rejection of Shanin's external leveling mechanisms). Thus peasant production is argued to be inherently unstable under the development of capitalist market relations.

This instability has been exacerbated in Latin America by the development of large-scale capitalist agricultural production. The development of this sector may affect the peasantry both through direct competition and through the development of a wage labor market. De Janvry presents the classic Marxist argument that competition from capitalist agriculture has destroyed the production role of the peasantry, both through direct product-market competition and through competition for land and other resources. "The peasantry grows in size but simultaneously loses its status as commodity producers. It is forced onto more and more minute and eroded plots, where it is of necessity becoming increasingly semiproletarianized" (de Janvry 1981 : 121, 172–73). This is the opposite of Chayanov's (1966 : 237) conclusion that the peasantry could drive up land rents to the point where capitalist production would be displaced. De Janvry's conclusion is based on the postulate of more efficient capitalist production (Patnaik's argument for different production functions in the two sectors).

Why, however, does the peasantry become "semiproletarianized" instead of fully proletarianized? Why is the peasantry not completely dispossessed by the more efficient capitalist units? De Janvry suggests an answer in his second level of analysis, which considers the particular characteristics of the agricultural wage-labor market as it has developed in the Latin American context. De Janvry argues that Latin American commercial agriculture needs "cheap labor" because of the nature of the region's overall economic development, which he characterizes as "disarticulated." Disarticulation refers to the focus of Latin American capitalist production (both agricultural and industrial) on domestic luxury and export markets, rather than mass-market production. As a result of this particular production bias, there is no "objective" (i.e., economic) pressure to create and expand the Latin American mass market by expanding the consumption capacity of the wage-labor force. There are, then, no objective factors creating a floor under the level of wages. At the same time, Latin American commercial agriculture is subject to a number of other economic and political pressures (expressed in the deteriorating terms of trade for this sector) that force it to seek means to keep costs down. The result is objective pressure to reduce real wages in the agricultural sector (de Janvry 1981: 26–37).

This provides the economic basis for semiproletarianization: peasants can supplement their wage income with use-value production on their small plots. This allows these semiproletarian workers to accept a wage below the level required by the fully proletarian worker (de Janvry 1981: 173).[24] This is similar to Bernstein's argument, discussed above, that peasant use-value production allows the African peasant household to accept a lower return on its cash-crop production—with one major difference. Bernstein still posits a commodity production role for the African peasantry under a process of capitalist vertical concentration of agriculture, whereas de Janvry posits prior destruction of peasant commodity production in Latin America by capitalist agricultural producers.

De Janvry's use of the term *semiproletarian* also differs from that of Amin. Whereas Amin's semiproletarian peasant earns the equivalent of a wage by working on the family farm and selling its pro-

24. "As they lose control over resources, peasants increasingly remain in production only to feed themselves—to complement the wage they receive for the labor they are now forced to sell to those who control the resources. Capitalist producers benefit from this semiproletarianization by paying a wage which is below the cost of maintenance and reproduction of the labor force, since the function of the peasant sector is to add the necessary complement" (de Janvry 1981: 173).

duce, de Janvry's semiproletarian peasant earns a wage by working elsewhere. Amin's peasant provides "cheap food" to the nonfarm sector; de Janvry's peasant provides "cheap labor" to capitalist employers. De Janvry argues that the proliferation of smallholdings in Latin America is evidence of this semiproletarian status of the agricultural labor force rather than of the persistence of peasant production.

Despite the functional role of the semiproletarian peasant sector to the capitalist agricultural sector—what de Janvry refers to as the "functional dualism" of Latin American agriculture—the destruction of peasant production by capitalist agriculture is so strong that the tendency is to "proletarianize the peasantry anyway" (de Janvry 1981 : 174). The resolution of this "contradiction" requires the intervention of the state, in the form of certain types of rural development programs that have as their purpose the perpetuation of the functional semiproletarian sector—those rural development projects classified by de Janvry (1981 : 230–31) as "economic" or "basic needs."

## An Assessment of the Current State of Latin American Peasant Production

De Janvry (1981 : 71–77, 109–22) provides considerable empirical support for the recent growth of large-scale commercial agriculture in Latin America and the concomitant increase in the number of smallholding households (generally, under five hectares). There is no doubt that many of these households combine farm income with off-farm sources of income, in most cases, agricultural wage labor. This is not sufficient evidence to argue, however, that peasant production for the market is being destroyed, that "peasants increasingly remain in production only to feed themselves," that "the middle peasantry disappears as a set of *producers* of commodities," and that "peasant foods cease to be produced as the upper peasantry is destroyed" (de Janvry 1981 : 172–73). As de Janvry himself demonstrates, peasant production remains an important component of total agricultural production in Latin America.

Much of de Janvry's argument about the "destruction of peasants as commodity producers" is couched in *relative* terms, as in the continuation of several of the above quotations: "the middle peasantry disappears . . . *relative* to the ever-increasing weight of capitalists in total production . . . their meager production becomes *relatively* scarcer," and "peasant foods cease to be provided . . . *to the extent* that they are still produced, the semiproletarians and the remaining middle peasantry produce them" (de Janvry 1981: 172–73, emphasis added). In very few instances, however, has peasant production actually declined over the past three decades. De Janvry's data (1981:72–73) show instead that peasant production has increased overall, although not as rapidly as large-scale commercial production or, in many cases, even as rapidly as population growth.[25] The CEPAL study cited above found that peasant production in Latin America "has an unarguable significance and importance," accounting in the early 1970s for 41 percent of all agricultural production for domestic consumption and 32 percent of all agricultural production for export (López Cordovez 1982:26, 27; Ortega 1982). Another CEPAL study found that overall peasant production had increased by an average annual rate of 3.4 percent between 1965–67 and 1975–77, comparable to the rate of growth of overall agricultural output (Ortega 1982:88).

In the case of Colombia, one of the three countries discussed in detail by de Janvry (1981:131–36) to illustrate the semiproletarianization process, peasant production has held its own as a share of total agricultural production despite the remarkable growth of large-scale commercial agriculture. One approach to the analysis of agricultural production in Colombia, initiated by Kalmanovitz (1974) and utilized by de Janvry, involves examining the behavior of groups of crops classified according to the types of units on which they are grown: "capitalist crops," which include cotton, refined sugar, rice, sorghum, and soybeans; "peasant crops," which include beans, plantains, cassava, and sugar cane for unrefined block sugar (*panela*); "mixed crops," including wheat, corn, and potatoes (grown on both large and small units); "plantation crops" (e.g., bananas); and coffee, grown on both large and small units.

While the "capitalist crops" increased from 10.6 percent of total agricultural output in 1950–55 to 30.8 percent in 1971–76, "peasant crops" remained virtually unchanged as a percentage of the

25. The peasant crop whose growth was weakest is wheat: affected by low-cost imports under the U.S. PL480 program, wheat production declined markedly in a number of Latin American countries.

total: 23.8 percent in 1950–55, 24.2 percent in 1971–76 (de Janvry 1981:132–33; Kalmanovitz 1978). This constant percentage share reflects an absolute increase in the level of production of peasant crops. De Janvry's conclusion (1981:134) that there has been a "decreased importance of crops produced by peasant producers" is true only in a limited and relative sense, with reference to the sharply decreased percentage share of "mixed" crops and coffee, produced on both large and small farms, in total agricultural output. This decrease does not necessarily reflect the destruction of peasant production by capitalist competition. In the case of coffee, the relative decline in production is at least in part attributable to the depressed market conditions for that crop in the 1960s. Looking at production by type of holding, rather than by crop, one Colombian government study found that peasant production contributed 63.2 percent of agricultural value added in 1973 for domestic foodstuffs, and 72 percent of production in 1976 of coffee, sugar cane, and cacao (Ortega 1982:85).

De Janvry's other argument for the destruction of peasant production in Colombia is that capitalist agricultural production is encroaching on peasant land. This is certainly the case with traditional estates where former tenant farmers have been displaced so that land could be turned over to "capitalist" crops (either directly or through cash rental arrangements with "capitalist" farmers). The sharp drop in sharecropping arrangements as a percentage of farm units under ten hectares between 1960 and 1970 is in part a reflection of this process,[26] and it has been documented in studies of the regions of Colombia undergoing most rapid development of the new commercial crops (Taussig 1974).

This is the "Junker" road to capitalist development described by de Janvry (1981:106–7). However, there is no evidence that small-scale owner-operators—the "family farmers"—are being displaced on the whole by this process. In the department (state) of Valle del Cauca, Colombia, a region that has experienced one of the most rapid transitions from traditional to modern estate production, units under ten hectares increased from 68 percent of the total in 1950 to 69.8 percent in 1960, and then declined slightly to 68.2 percent in

26. Sharecropped units under ten hectares decreased from 125,178 in 1960 (13.5 percent of all units under ten hectares) to 57,843 in 1970 (6.7 percent of all units under ten hectares). (Calculated from DANE 1973a:26, 29, tables 5 and 10.)

1970 (DANE 1954 and 1973a). A sharp drop in small-scale sharecropping was countered by an increase in the percentage and number of small-scale owner-operated units.

The same was true in Colombia as a whole. The total number of units under ten hectares and the acreage controlled by those units decreased between 1960 and 1970 from 925,750 units and 2,403,725 hectares, to 859,884 units and 2,234,283 hectares. In 1970, holdings under ten hectares in Colombia accounted for 73 percent of all agricultural units and 7.2 percent of all land in units, compared to 76.5 percent of units and 8.8 percent of land in 1960 (DANE 1973a:26, tables 5 and 6). However, this decline was accounted for by the sharp decrease in small-scale sharecropping, while during this same decade, the number of owner-operated units under ten hectares had increased (from 554,581 to 579,295 units), as had their acreage (from 1,475,748 hectares to 1,497,670) (DANE 1973a:29, table 9). For Latin America overall, de Janvry himself points out that "in the majority of countries, the growing number of peasant farms led to an increase in both the percentage of farm holdings and the percentage of total agricultural land controlled by peasants, even though here there are more exceptions" (de Janvry 1981:121). CEPAL researchers have found that for eight Latin American countries, including Brazil and Colombia, the percentage of units under 20 hectares (a more expansive definition of peasant units) increased between 1960 and 1970, as did the land they controlled, both absolutely and as a percentage of all land in farms (Ortega 1982:99–101).

There is, then, very little empirical evidence for the "destruction" of Latin American peasant production in terms of land farmed or output, despite the very rapid growth of capitalist agriculture. This is not to say that "semiproletarian" units, defined as agricultural units that provide only a portion of overall household subsistence needs, have not increased in number. They have, and their size has decreased due to fragmentation. Nevertheless, it is important to distinguish between the commodity production role of these units in the overall structure of the economy and their reproduction role in providing sustenance for the farm household. While the importance of commodity production in overall household income may have diminished, as de Janvry argues, this is not evidence for the decline of peasant commodity production in the economy as a whole.

A key feature of Latin American agriculture is that it is still possible to classify many products as either "peasant" or "capitalist." Traditional Latin American staples continue to be grown primarily on smallholdings (López Cordovez 1982:26, table 6; Ortega 1982:84–85), while capitalist agriculture has concentrated on products for export, for domestic industry, and for domestic higher-income markets. Only a few traditionally "peasant" crops have been heavily displaced by imports and/or domestic capitalist production (e.g., wheat and rice).

This "division of labor by crop" is reflected in the different behavior of output and prices in the "capitalist" and "peasant" agricultural sectors. The capitalist crops have been marked by rapid output growth and declining real prices. De Janvry (1981:169) argues that this is an inherent characteristic of the capitalist sector, since "capitalist production is synonymous with accumulation and growth." Conversely, "those crops which continue to be produced on a peasant basis show relatively stagnant production trends and increasing prices within the margin allowed by cheap food policies." De Janvry (1981:73–74) provides evidence for corn, beans, and cassava (manioc), showing that "the prices of crops whose production rates were most stagnant (e.g., manioc) nearly doubled between 1950 and 1975."

This behavior of output and prices in the "peasant" sector is not what we would see in the classic process of destruction of peasant agriculture envisioned by Patnaik, which involves direct competition by more efficient capitalist units, the decline of peasant production, and yet stable or declining prices as the capitalist units expand their output. It is consistent with adjustment of peasant production to meet the combined effects of diminishing returns (due to population growth, fragmentation of holdings, and soil erosion and exhaustion) and growing wage-labor opportunities. As capitalist agricultural production develops in some crops, the growing demand for wage labor may draw some peasants out of commodity production—particularly if the combined pressures of vertical concentration and competition between peasant producers have driven the returns to peasant farm labor below the capitalist wage. As some peasants leave farming, however, the decreasing relative supply of peasant products drives their prices up until returns to farm labor equalize with returns to capitalist wage labor (minus, perhaps, a nonpecuniary gap). This is the observed pattern for peasant crops in Latin America.

Since capitalist agriculture does not set the level of agricultural supply prices for "peasant" crops, a stable division of commodity production between capitalist and peasant farms can be maintained even in the face of a growing wage-labor market, through adjustments in the return to farm labor. However, this equalization process can be hindered by government efforts to prevent staple food prices from rising, as de Janvry argues has been the case in Latin America. In this case, we could continue to see returns to peasant farm labor considerably below the level of wages, despite the shift of some peasant labor from commodity production to wage employment.[27] This is not evidence that peasant commodity production is no longer technically competitive, however, as de Janvry implies.

In sum, although capitalist agriculture has developed rapidly in Latin America, it has not yet moved into a number of spheres that continue to be dominated by peasant production. Peasant commodity production persists, along with the involvement of some household members in various off-farm economic activities, including agricultural wage labor. In the absence or low level of development of capitalist production in many traditionally "peasant" crops, the levels of output and prices of those crops have responded primarily to conditions of peasant production relative to demand.

## Conclusion

The conclusion of this review is that we cannot yet dismiss the role of the Latin American peasantry as agricultural commodity pro-

27. Equalization of farm returns with wages can also be inhibited if nonfarm income-earning opportunities have not expanded sufficiently, so that peasant commodity prices remain depressed due to "too much" production relative to "costs" of production (necessary household income). Even in the context of rapid expansion of agricultural wage labor, labor market limitations frequently persist with respect to rural wage-labor opportunities for women and children. This may lead to an age- and gender-specific division of labor within smallholding households, with male members working primarily off-farm while women and children engage in both commodity and use-value production on the holdings. The zero or

ducers. This is not to assert that the behavioral characteristics of peasant producers imply their universal superiority over capitalist agriculture, as the Chayanovians argue. Capitalist agriculture is clearly developing in Latin America, making inroads into some products formerly dominated by peasant production, and also absorbing growing numbers of peasants as wage laborers. Patnaik is correct in pointing to the role of technological change in the development of increasing productivity on capitalist farms. I do, however, reject the classic Marxist argument for the universal superiority of capitalist agriculture over family-farm production. To date, peasant production in Latin America has persisted in some crops along the lines of the vertical concentration path postulated by some Chayanovians, although with an important element of "semi-proletarianization" in the sense of combining peasant commodity production with wage labor.

This may well be, as the classic Marxist position claims, only a transitional stage in an inexorable process of total dispossession of family farming. Yet this process is operating, at best, in an extremely slow and uneven manner, and this slowness and unevenness should be the precise object of our analysis. If we reject both determinist positions on the inevitable superiority of one or the other form of agricultural production, we are led to focus instead on the reasons for the unevenness of capitalist agricultural development and the resulting persistence of economically competitive peasant production. The rest of this book considers this question in the context of a study of family farming in an Andean region of southwestern Colombia.

---

low opportunity cost of this family labor may lead to its application even when the return is considerably below the level of (adult male) agricultural wages.

*Chapter Two*

# The Establishment of Family Farming in a Colombian Mountain Region: Fifteenth to Early Twentieth Centuries

The factors that have affected the development of family farming in Colombia can be seen more clearly by examining this process in concrete instances. This book focuses on one such instance: the community of El Palmar, in the *municipio* (county) of Dagua, department of Valle del Cauca. How family farming came to be established in this region of Colombia is an important question not only for its historical significance, but also because an examination of the roots of peasant agriculture will provide us with a basis for analyzing subsequent patterns of agrarian change. In considering the emergence of family farming in the Dagua region, I shall relate the changing patterns of regional economic development to the key conditions of access to land, to labor, and to markets.

I shall first trace the development of settlement patterns, transportation networks, land tenure, and economic activity in colonial Valle del Cauca, to define the context in which the settlement of the community of El Palmar took place in the late 1800s. The reader is encouraged to familiarize him/herself at this point with the landmarks depicted in map 1: the settlements of Cali, Popayán, Buenaventura, Dagua, El Salado; the rivers of Cauca, Dagua, Digua, and Anchicayá; the western range of the Andes; the mining districts of

Raposo and Chocó; and the general location of Valle with respect to Antioquia, Bogotá, and Cartagena.

## The Colonial Period: Turning Inward

The basic outlines of the economic structure of Valle del Cauca were laid during the colonial period. Between 1535 and 1540 three Spanish expeditions made their way inland from the Pacific coast, through dense lowland jungle, towards the interior of southern Colombia. Two expeditions entered along the rivers Anchicayá and Digua. One reached the valley of El Salado, an area of some agricultural promise. The other reached the Raposo region, which was found to have considerable potential as a mining district. The third expedition entered by the Dagua River, eventually crossed the Western Andes (Cordillera Occidental), and reached the Cauca River Valley, where the settlement of Cali was founded.

This rich, fertile valley, with its pleasant climate, had already been explored by expeditions that arrived overland from the south and the north. The settlement of Cali, founded in 1536, was quickly surrounded by immense estates (latifundios) granted to the early Spanish conquerors. These settlers were interested in establishing a trade route to the Pacific. However, the area between the coast and the valley was populated by scattered groups of Indians, who mounted fierce resistance to the Spanish. Early attempts to establish a settlement on the coast and an outlet from Cali to the ocean failed because of Indian attacks. The coastal town of Buenaventura, founded in 1539, was burned to the ground in 1597 in one such conflict (Gomez Benítez 1979).

Despite pressure from the settlers, Spain did not support the establishment of a Pacific port. The region was difficult to defend because of its sparse population. It was more economical to concentrate trade through one port because of the cost of protecting against pirate raids, controlling gold export traffic, and establishing customs facilities (Gomez Benítez 1979). All trade was therefore channeled through the Caribbean port of Cartagena.

This implied enormous overland transportation costs for the settlers of southwestern Colombia, who therefore defied Spain and attempted throughout the colonial period to establish an outlet to

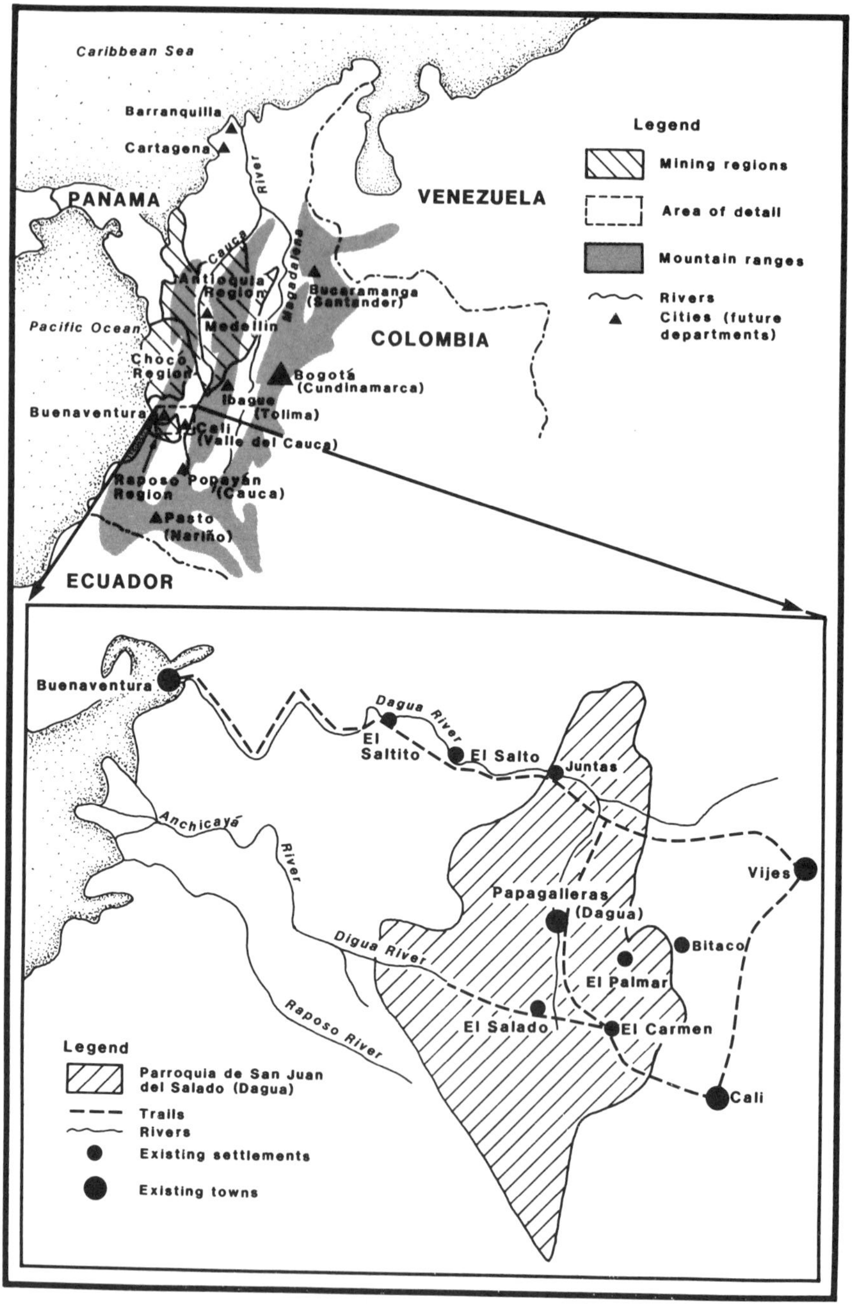


Map 1. Early Nineteenth Century: Major Mining Regions, and Detail of Cali-Buenaventura Region.

the Pacific. The major route from Cali followed the path of the original Pacific expedition, along a trail used by the Indians in their periodic excursions to the ocean for salt (Franco Mejía 1980; see the northern route in map 1). It was a long and difficult journey, which culminated in the navigation by canoe of the lower reaches of the Dagua River as the only means of passing through the impenetrable jungle region. Despite the potential value of this route to the development of the Cauca Valley region, the difficult topographical conditions and lack of Spanish government support posed obstacles to any improvements. At the end of the colonial period, trade from the valley was still being carried out by arduous mule trails and treacherous canoe navigation.[1]

It has frequently been noted that the failure to develop the Pacific route during the colonial period limited the economic development of Valle. With the exception of trade with the Chocó mining region to the north and some export of cattle and mules to the south, Valle remained geographically and economically isolated despite its considerable agricultural, mining, and commercial potential (Franco Mejía 1980).[2] It was forced to turn inward.

A little-noted result of this inward focus was the encouragement

1. Many attempts were made to improve the Dagua route. As early as 1564, the governor of Popayán issued a proposal to establish trade along this route. In 1582 a license was given to one Francisco Jaramillo to open a road by this route, which was a footpath impassable by horse traffic. The trail was left in rudimentary form when the project failed. By the end of the colonial period, the path had been improved to a difficult mule trail. The trip from Cali to Buenaventura, a distance of about 120 kilometers, took three days. Nevertheless, this route carried most of the external trade of the southwest, including Popayán (Ortega 1920:451–52; Franco Mejía 1980).

2. A comparable process was taking place throughout the region that was to become Colombia. By the end of the colonial period, this portion of the Spanish holdings in South America consisted of five distinct regions: the southwest (later Chocó, Valle del Cauca, Cauca, and Nariño); the northwest (Antioquia); the coastal lowlands (later Atlántico, Magdalena, and Bolívar); the central highlands (later Boyacá and Cundinamarca); and the north-central region (later Santander and Norte de Santander). These regions were isolated from one another as a result of Spanish colonial policy, and had developed distinct economic and social structures due to differences in climate, geography, and labor force availability. While there was some interregional trade based on these differences, particularly in textiles, salted meat, cattle, sugar, salt, and cocoa, this trade was limited by the poorly developed transportation routes. For descriptions of each of these regions, see Mejía 1971:50–51; Escorcia 1978:14–17; Ospina Vásquez 1955:19–34, 52–74; Parsons 1968:2, 65–66; and Nieto Arteta 1975:57–59.

it gave to the development of a differentiated internal economy and intraregional trade. The basic outlines of this development, with the varied patterns of land tenure and land use to which it gave rise, can be reconstructed by piecing together evidence from various sources.[3] In so doing, we find that the region was far from stagnant. Although also far from fulfilling its potential, the economy was nevertheless expanding and changing during this period in a manner that laid the basis for future growth.

## *The Haciendas*

In the sixteenth and seventeenth centuries, as a result of the scarcity of an indigenous labor force in the valley, the land of the latifundios around Cali was used primarily for extensive cattle production. In this period, the Valle area was of little economic importance, and Popayán was the major seat of colonial power in the southwestern region (Franco Mejía 1980). By the late seventeenth century, the development of the Chocó and Raposo mining districts gave new life to the region, and the locus of economic and political power in the southwest began to shift from Popayán to Cali.

By the eighteenth century, the valley lands to the west of Cali were beginning to be concentrated in the hands of mine owners, particularly from the Raposo region, and the growing merchant class (Colmenares 1976:55–67, 214–15, 228). These lands, which had been gradually fragmented through inheritance in the sixteenth and seventeenth centuries, began to be reconstituted into what Colmenares terms *haciendas*, as opposed to the more traditional latifundios, which still predominated to the east of the Cauca River. In addition to cattle, the haciendas produced a variety of

3. Work is just beginning on the important task of reconstructing the economic history of this region of Colombia. Most of the available research focuses on the nineteenth century, with the notable exception of Colmenares' valuable work on eighteenth-century haciendas (1976). Other important works are Escorcia 1983; the useful theses on aspects of Valle history by Hyland (1979), Gomez Benítez (1979), and Franco Mejía (1980); Banderas' detailed history of communities and transportation routes (1944); and the section of the first volume of Ortega's railroad history dealing with the Cali-Buenaventura line (1920). I have used these sources in conjunction with population censuses, church records, and oral histories to trace the outlines of the emergence of family-farm communities in this region.

products for the mining districts, especially the sugar-cane liquor *aguardiente* (Colmenares 1976:102–6).[4] The basis of this expanded agricultural activity was the ability of the miners and merchants, drawing on capital from their other ventures, to invest in slaves to work the hacienda land. In fact, Colmenares suggests that an important impetus to these land purchases was the ability of mine owners to transfer slaves from their mines to the haciendas (Colmenares 1976:96–102).

There is also evidence of hacienda formation in the mountain areas west of Cali by the early eighteenth century. Nicolas de Caicedo Hinestroza, for example, who owned mines in the Anchicayá River and Dagua River regions, also had vast landholdings in the mountain region, including the area to the north of El Palmar (Bitaco) and the area now known as Dagua (then Papagalleros). Hinestroza's lands reached in the north at least to Juntas (now Cisneros), and in the south to the border of El Salado and into the Raposo mining district (Colmenares 1976:61, 251–53). By 1760 part of this land, including holdings in the Raposo district, was acquired by the mine owner Guillermo de Collazos y Ayala and his brothers, who also owned much of the land in the El Salado region. Their extensive landholdings reached to the Las Cruces hill west of Cali, and included within its boundaries the *camino antiguo de Dagua* (the original trail through Dagua from Cali, shown in map 1) (Colmenares 1976:61, 256).[5] Another Raposo mine owner, Nicolas Perez Serrano, also owned a hacienda in the Dagua region (Colmenares 1976:99).[6]

In 1774, when one of the Guillermo Collazos daughters inherited part of the El Salado holdings, her inheritance included a

4. This was particularly important during the first half of the eighteenth century, when *aguardiente* was sold to the slaves in the mining districts. After 1758 this was prohibited by law, and after 1763 attempts in Cali to enforce the tax on *aguardiente* that had been established earlier in the century further decreased *aguardiente* production.

5. The *camino antiguo* was later superseded by the northern trail shown in map 1, according to Franco Mejía (1980). Nevertheless, it is clear that this route remained an important link between Cali and the El Salado region, and ultimately the Dagua region, as we shall see.

6. Escorcia (1983) refers to the haciendas of Platanares, El Salado, and Dagua, which would be the remains of these three estates.

*trapiche* (a mill for extracting juice from sugar cane), cultivated land, one hundred beef cattle, thirty horses, and fifteen slaves (Colmenares 1976:257). This was apparently a diversified agricultural enterprise. It is likely that her father had been using slaves from his holdings in the Raposo mining district to work his hacienda in El Salado, thus providing foodstuffs for his mining population. Colmenares argues that the mine owners' greatest preoccupation was feeding the slaves of the mining district. The mining rights included surrounding cultivable land, and Colmenares found frequent mention in the land sale records of plots of plantains and maize. Juan Francisco Garcés, for example, grew plantains, maize, and cacao in his mine lands in the Dagua region (Colmenares 1976:152–53).

Thus mining and the agricultural development of the western region went hand in hand. Mine owners acquired vast landholdings in the mountain areas, which they worked with slaves from that same mining population. But this production, according to Colmenares, was insufficient, and the mining population also depended on the agricultural production of the valley haciendas.

## *Smallholding Production*

The basic outlines of the development of the slave haciendas, both in the valley and in the mountain zone, are now evident to us thanks to the painstaking reconstruction of land transaction records by Colmenares. However, this is only part of the story. The other part, the development of smallholding production in the valley and the mountains, is more difficult to reconstruct, since these settlers generally farmed their lands without formal title to the property.[7] Nevertheless, there are some clues that this process was taking place.

The 1535 expedition that reached the valley of El Salado encountered a potentially rich agricultural district and founded the settlement of El Salado. This became the government seat (*cabecera*) of

7. The notary archives for the region, housed in the Archivos Históricos in Cali, may contain some information on early land transactions by family farmers. As of January 1984, however, these records were uncatalogued. Their systematic analysis is a project well worth undertaking.

the district of El Salado, and apparently prospered. By 1784 it had a church,[8] and the area that later became the county of Dagua was known as the Parroquia of San Juan of El Salado. In 1795 the town hall was built (Banderas 1944:136–37). The development of this settlement owed much to its favorable location between Cali, Raposo, and the Dagua River. It had access by trail to Cali and, along the *camino antiguo*, to the Dagua River route to the Pacific. It also had access, by the original Spanish expedition route, to the Raposo mining district. It was thus a strategically located settlement made even more attractive by its favorable climate and rich lands.

Evidence that the mountain region was not totally under the control of absentee-owned slave haciendas is provided by the church records of the Parroquia of El Salado. According to these records, there were twenty-two baptisms in 1810. In the periods from 1810 to 1823 and 1828 to 1837, the number of recorded baptisms averaged at least seventeen per year.[9] The names suggest that these were children of *mestizo* (of mixed Spanish and Indian ancestry) settlers who had moved up from the regions of Pasto and Popayán to the south.[10] These migrants apparently settled in areas along the two major trails through the region, the northern route

8. The church was built by the miner Andrés Guillermo de Collazos on his land (Colmenares 1976:77).

9. These records were located in the church in Dagua. They provide a rich source of demographic information for the district, going back to the early 1800s. There is information on births, deaths, and marriages (none of which, however, forms a complete series for the period). Sadly, the books are deteriorating and some of the information is already lost. No data or only partial data were available for 1824 through 1827. In addition, some data may be missing for the years of 1828, 1830, and 1831, so that the total for the period may be understated. Census data on population growth in the Valle del Cauca region and in the nation as a whole are reviewed in Appendix A.

10. Although discussions of Valle mountain settlement generally emphasize the movement south from Antioquia, the formal El Palmar interviews and informal discussions with residents revealed that the population of much of this area today is descended from migrants from the southern area. This information will be presented in more detail for El Palmar. The names in the baptismal records are the same ones encountered in the population censuses today. The majority of these descendants of the early settlers are mestizos, with scattered areas, as in the town of Dagua itself, where one finds some concentration of descendants of the slave population. This group of free black settlers became increasingly important after independence, with the manumission of slaves, and particularly after the abolition of slavery in 1851.

through the mountains to the Dagua River, and the southern route through El Salado to the Raposo mining district. It is unclear whether any settlement took place prior to the 1800s in the interior region between these two routes, along the *camino antiguo*. For this perod there is also little information on smallholding agricultural production, although we can speculate that some production may have supplied the mining districts and perhaps even the city of Cali, which by 1787 had 5,690 residents (DANE 1955:19). In the later nineteenth century, the picture becomes clearer, as we shall see.

In the valley lands, also, smaller holdings were emerging alongside the latifundios and large haciendas (*haciendas mayores*). These ranged from the smaller haciendas (*haciendas menores*), down through medium and small properties. These smaller holdings grew foodstuffs such as plantains, maize, rice, and beans for the slaves of the mines and the growing population of Cali (Colmenares 1976:106–7, 184–88, 205–63; Escorcia 1980:56–57). However, the monopolization by the larger estates of the valley lands around Cali probably slowed the process of independent smallholder formation. In the 1800s this led to growing foodstuff shortages in Cali and an increasing dependence on the agricultural production of the mountain holdings to supply that city's food needs.

## The Nineteenth Century: Consolidation of a Mountain Economy

Although the development of Valle during the colonial period was limited, by the end of this period there are clear signs of the emergence of a differentiated economy, which included the slave mining sector, the cattle latifundios, the slave haciendas, the food-producing smallholding sector, and the urban sector. Despite poor internal transportation routes, these sectors were already linked by agricultural trade. This trade continued to develop in the nineteenth century as the population of the region grew. In this period we see the consolidation of the mountain economy, linked to the food needs of the urban sector, to the commercial/transportation expansion of the region, and to the particular forces operating within the smallholding sector. Although much has been written about the nineteenth-century colonization processes in Colombia that were linked to external trade, the type of internal development that occurred in Valle, which is equally important, has not been examined in the same depth.

## *Migration to Dagua: Origins of Settlers*

Although much of the nonslave population of Valle lived in the towns of Cali and Buenaventura, some rural settlement was taking place. The population of the mountain region between these towns slowly, but steadily, increased during the nineteenth century through immigration and natural increase. The settlement of this mountain region owed much to a previously undocumented migration of mestizo farmers coming north from southwestern Colombia. Of the nine men interviewed who had arrived in El Palmar between 1890 and 1930, five came from rural areas of Valle, Cauca, or Nariño. Before looking at their experiences upon arriving in this region, let us consider the conditions in their regions of origin and the factors that led them to migrate.

According to one of the early settlers, Juan Álvarez, many of the settlers "came from Cauca because in Cauca they were all very poor, the farms were small and the land sterile and there was no land left." The southern provinces of Cauca and Nariño were more densely populated than the mountains of Valle. The Spaniards had encountered larger and more easily subdued Indian populations in these regions. Areas around Popayán and Pasto were carved into vast latifundios and smaller *resguardos* (Indian reservations), with Indian and slave labor used to work the estates. As in other mountain zones of Colombia with similar conditions, such as the plateaus of Boyacá and Cundinamarca, the later Spanish settlers ("poor whites") and the mestizo population established smallholdings next to the large estates (Escorcia 1978:45; Fals Borda 1957:84). There is some evidence that in the colonial period and in the early 1800s, these were areas where family farms produced foodstuffs for sale to merchants, urban areas, and local markets, as well as for family consumption (Melo 1978:45).

The mestizo population grew rapidly, soon pressing on the concentrated landholdings of the estates. The subdivision of holdings through inheritance, which had begun in the colonial period after the elimination of primogeniture, resulted in the fragmentation of the family farms in the highland regions and the development of more complex livelihood strategies. The histories of the families of some of the El Palmar settlers illustrate the ways in which these mestizo households earned a living by the late nineteenth century.[11]

Aristides Benavides was born in 1907 in La Cruz, Nariño. When he was young, his family left La Cruz because it was difficult to find land or work. They moved to Tambo, Cauca (near Popayán), where his parents rented land from a hacienda in exchange for days of labor on the estate. They grew maize, potatoes, wheat, and onions, which they sold in Tambo and Popayán. After five years their earnings enabled them to acquire land of their own, thirty acres in cattle and crops. Market conditions and usufruct arrangements were thus favorable enough to allow some landless households to acquire land and to establish themselves as independent family farmers in this region.

Aristides himself followed another route to land acquisition in the area: migrant agricultural labor. At the age of ten he left with a man from Huila, who took him to San Agustín. There he worked for ten to twenty centavos a day on a hacienda. After a year he went back to La Cruz, then returned to Tambo, where he was able to buy some land with his earnings. By the time he was thirteen, Aristides had cattle and crops.

The family of two other migrants to El Palmar, Lubino Cabrera and Agapito Cabrera, also resided in southwestern Colombia. Lubino was born in the mountains near Pasto, Nariño, in 1895. His father had a farm in the warmer zone, where they cultivated basic foodstuffs. When Lubino was a young man, the oldest son of eight children, his father decided to transport sombreros from the region of Tumba to sell in Guayaquil, Ecuador. He made several trips, accompanied by Lubino and another son, but the journey was difficult and the danger of robbers great. Lubino's father decided not to continue with the business, and they stayed in Guayaquil. There they rented land from a rice and cacao hacienda, growing food crops in return for days of labor on the estate. Lubino met and married his wife Ursulina, and they had a child and returned to Nariño. There they lived with Lubino's brother Agapito, who had stayed

11. These histories are based on interviews with Aristides Benavides and Ursulina and Lubino Cabrera in 1978–79 and 1983–84.

behind and acquired a farm. Their parents returned from Guayaquil soon after, having saved enough to purchase land in the cooler zone, and Lubino worked *en companía* on his father's farm growing wheat, potatoes, maize, and beans. *En companía* is a type of sharecropping arrangement common to Colombia's mountain zones. It is frequently entered into between family members, particularly between older smallholders and their landless younger relatives. Traditionally, the landowner provided the land and perhaps some of the inputs other than labor, and the output was shared fifty-fifty.

Why would young men such as Aristides, Agapito, or Lubino subsequently migrate, leaving behind family and friends and a degree of economic security to strike out on their own in unknown territory? Limited economic opportunity in the region of origin is commonly considered to be one of the principal motives behind the nineteenth-century migration out of the regions of colonial settlement, such as the Popayán area. However, the histories of these migrants to El Palmar show that the process was often more complex. Migrants were not necessarily landless peasants; in some cases their parents had land, while in other cases they themselves did. Market production through land rental, agricultural day labor, and trade had provided the means for some land-poor households to acquire land even in the more densely populated southern regions. The migration of young men from these older smallholding communities to incipient ones in the Dagua mountains was a result not simply of relative economic opportunities in the two regions, but also of intrahousehold relations on the family farm.

These relations were far from egalitarian. The male head of household was the ultimate authority in all decisions regarding the farm and many decisions regarding the family as a social and cultural unit. Other family members received the fruits of the family's collective farm labor only at his discretion. He generally carried out all sales of farm products and allocated the proceeds between farm investments, expenditures for family needs, and possibly savings. Prevailing norms supported an unequal distribution of benefits among family members, with women and children in particular having little recourse even when his decision meant their material deprivation.

Recent research from rural communities around the world has documented systematic age- and gender-related inequalities between family members with respect to nutritional status, education, leisure time, and other measures of social and material status, with elder males being most favored (Caldwell 1982: 165–66; den Hartog 1973; Kynch and Sen 1983; Bourque and Warren 1981; Anker, Buvinic, and Youssef 1982; Hirschon 1983; Mosk 1983; Parsons 1984; Henn 1986 and 1987; Folbre 1985 and 1986; Kabeer 1985). In their efforts to understand the bases for these differences, numerous scholars have postulated various forms of "patriarchy" as the explanation. Although the term *patriarch* has long been used in a descriptive sense to refer to a male head of household, feminist writers in the 1970s began to give this term a more theoretical content. As Beechey (1979:66) expressed it, "the theory of patriarchy attempts to penetrate beneath the particular experiences and manifestations of women's oppression and to formulate some coherent theory of the basis of subordination which underlies them." The application of the theory of patriarchy to the analysis of unequal relations in the traditional farm economy has its roots in the general theoretical development of this concept by feminist scholars.

While sharing a common understanding of patriarchy as a system of elder male domination over women and younger males, feminists have been divided over its theoretical content. What are the roots of patriarchy? In particular, what is the relative importance of biology, ideology, and economy as explanations of women's subordination? How does the development of capitalism affect patriarchal relations, and vice versa? Most writers have emphasized the importance of women's childbearing potential (their biological reproductive capacity) as an incentive for male domination and control.[12] Some writers have also pointed to the importance of women's productive potential as an incentive for male domination (Delphy 1977). This includes women's domestic work, not only childrearing, but also labor that provides the male with meals, clean clothes, a clean environment, and other services; as well as women's market work.

12. The interpretation of this factor varies widely, however, ranging from biological determinism (Brenner and Ramas 1984; see the review of "revolutionary feminist" writings in Beechey 1979:69–70) to nuanced analyses of the relationship between women's childbearing potential and the social construction of gender and labor force (Rubin 1975; Edholm, Harris, and Young 1977; Eisenstein 1979; Henn 1986; Stichter and Parpart 1987).

The relative weight of these factors varies widely in different economic contexts, for example in a traditional farm community versus an advanced industrial society. Much feminist literature has been concerned with understanding why and how patriarchy persists under advanced capitalism, which has eroded children's economic contribution to the household and replaced much traditional women's work in the home by industrial goods and services. Here debates have raged over the importance to patriarchal persistence of the continued socialization of children into gender roles that have been rendered obsolete by changing material conditions (ideological autonomy: Mitchell 1974), the importance of continued male benefits from women's subordination (Hartmann 1981a and 1981b; Eisenstein 1979), and postulates of a real economic (or "materialist") base for patriarchy under capitalism. Central to this latter focus has been the "domestic labor debate" among Marxist scholars over the role of women's unpaid domestic labor in subsidizing the husband's income and thereby enabling his employers to pay him a lower wage (Seccombe 1974; Gardiner 1975; Himmelweit and Mohun 1977; Molyneux 1979; Folbre 1982; Bryceson 1983; Henn 1986). Other Marxists have pointed to the effect of women's domestic work in cheapening their own wage labor (Beechey 1977). In the developing countries as well, Marxists have argued, women's production—here including not only domestic work but also all the forms of women's agricultural and petty commodity production that create a "semiproletarian" labor force—subsidizes male workers and thereby lowers the wage (Deere 1976; Deere and de Janvry 1979).

Scholars have also drawn on historical analyses to examine the formation and functioning of patriarchal relations in traditional economies, as well as to explore the complex interaction between these traditional relations and the developing capitalist economy (Stichter 1975–76 and 1985; Guyer 1980; Henn 1986; Deere 1988; Caldwell 1982; Mosk 1983; Folbre 1983; Beneria and Sen 1981 and 1982; Meillassoux 1981; Bourque and Warren 1981). While debates continue over the relative importance of ideology versus economics to the explanation of patriarchal relations in the advanced industrial countries, a materialist explanation is more readily apparent in the traditional rural household economy with its central reliance on the family labor force (although the importance of biology versus economy in explaining women's subordination remains a contentious issue).

In the rural communities of nineteenth-century Colombia, the basis for the favored material status of the patriarch within the household was his control over the labor of subordinate family members. His status in the community depended on his ability to deploy that labor to accumulate and develop other economic assets, primarily land. As farm manager, the patriarch continually pressed household members to intensify their labor effort. As head of an economic unit dependent on a family labor force, the patriarch desired high fertility and therefore control over the reproductive potential of his wife. These aspects of traditional patriarchal relations will be illustrated in the early history of El Palmar.

How was compliance with patriarchal goals obtained from subordinate household members? One important factor was ideology. To reduce conflict, children were taught to accept their role as laborers unquestioningly—even willingly. Aristides Benavides was proud of the fact that he had been drilled in the work ethic: "In that period [when he was a child], parents took on the responsibility of teaching the child from a very young age to work." A daughter of another early settler, Juan Álvarez, also recalled with some pride the hard work she had done as a child, even carrying heavy loads of sugar cane on her shoulders over mountain trails. These norms, passed on from parents to children, helped to elicit compliance with the hierarchical structure of the family unit and the often onerous workload imposed on the offspring.

Yet socialization into one's age and gender roles is rarely enough to eliminate the tensions and conflicts inherent in the patriarch's role as manager and ultimate authority figure. Sons are being trained as the new generation of patriarchs, with all that that implies for rejecting a subservient role. Daughters, witnessing the toll on their mother's health of a heavy burden of work and continual childbearing, might wish to transform or reject that fate for themselves. Wives, too, may well chafe under their burdens. The patriarch's domestic authority depends crucially on broader economic, legal, political, and cultural practices that reinforce children's and women's subordinate position within the household by limiting their access to productive resources, to cash employment opportunities, and to credit to start up their own businesses, or by restricting their control over earned income (Afshar 1985; Standing 1985; Henn 1986; Folbre 1984).

In nineteenth-century Colombia, traditional supports for patriarchal dominance over sons were weakened by the attractive option of frontier settlement. In the context of these external opportunities, the interaction between father and sons and the training the sons received as incipient patriarchs played themselves out in an expressed wanderlust on the part of the sons. When possible, they left at an early age, even in cases where the farm provided an adequate standard of living for the family. Aristides left home for the first time at the age of ten: "I left on the sly. I was sent to do an errand and I didn't return." When asked why he left, he replied, "Because I liked to wander and wander." This motive, "to see the world," was cited as their principal reason for leaving home by four of the twelve migrants interviewed in El Palmar who had arrived in the community before 1940. Daughters also were anxious to leave the patriarchal household. However, there is no record of women arriving in the El Palmar area except as wives or daughters in patriarchal households. Independent migration to frontier areas appears to have been a male prerogative.[13]

The patriarch in nineteenth-century Colombia had no means of preventing his sons from leaving. He could encourage them to stay by sharing some of his economic power with them: paying them for their labor input or giving them land of their own to work. However, the ability of the patriarch to make an attractive offer in the areas of colonial settlement diminished as the farms became smaller over time. Even more fundamentally, the relinquishing of power in this way undermined the traditional relations of patriarchy, and was probably resisted by the patriarch even when he had sufficient resources at his disposal.

The "push" of diminishing economic opportunities was clearly an important factor in the nineteenth-century migration in southwestern Colombia: seven of the early migrants to El Palmar cited lack of land and employment as their principal motive. To this contributing factor, however, we must add the "push" of patriarchal relations. Even in the case of those leaving for ostensibly economic

13. Females may have had the possibility of migrating independently to the cities, a phenomenon for which there is some evidence in the nineteenth century for a number of Latin American countries (Kuznesof 1980: 102).

motivations, the "lack of land and employment" may have had its roots at least partially in familial, as compared to broader social, relations. Despite the conflicts inherent in the patriarchal relations, however—conflicts that had impelled many of these men to leave home—the migrants would attempt to establish similar patriarchal farm communities at their destinations.

## *The Settlement of the Dagua Region*

*The Álvarez Family.* Juan Álvarez, born in 1889, came from Popayán to the Dagua region with his parents in 1897.[14] They settled first in Olaya Herrera, where there were a few scattered farms (see map 2). After the War of a Thousand Days (1899 to 1902), in which his father fought, they moved up the mountain to an area that was to become the community of La Virgen, neighbor community to El Palmar.

All the land there was *baldíos*, public land free for the claiming, and they claimed their lot. They began by cultivating maize and beans on a small plot. After a few years they started to grow sugar cane and built a *trapiche* to extract the juice from the cane and cook it down into molasses (*miel*) and block brown sugar (*panela*), both of which they sold. The *trapiche*, which consisted of hand- or horse-drawn rotating wooden wheels to crush the cane and copper kettles to cook the syrup, represented a considerable investment, which would be undertaken only by farmers who had sufficient land in cane and good enough market prospects to justify the expenditure. Juan's father also began to produce and sell *aguardiente*, an illegal practice (the state monopoly of *aguardiente* was a major source of government revenue) for which he was arrested and served a year in the Cali prison. By this time Juan's mother had died, leaving behind six children. The boys worked as agricultural day laborers (*jornaleros*) for other farmers in the area.

Juan began slowly to acquire land of his own, some *baldíos*, but mainly improved land (*mejoras*) purchased from other settlers with his *jornalero* earnings. To cultivate his growing holdings, Juan used a system of *roza*, cutting down a section of trees and undergrowth,

14. This history is based on interviews with Juan Álvarez in 1978–79 and with members of his family in 1978–79 and 1983–84.

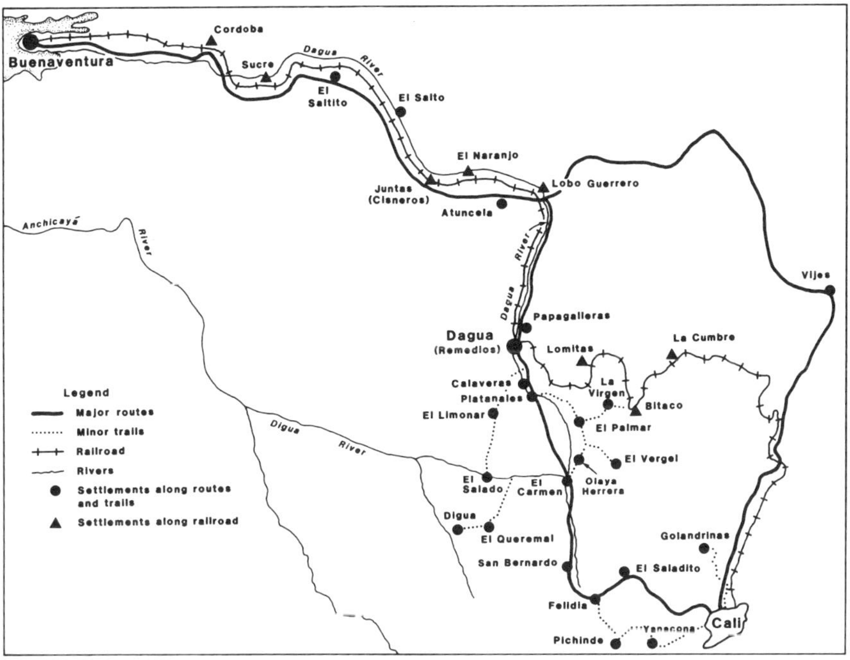


Map 2. Settlements and Transportation Routes in the Dagua Mountains by 1920.

burning it, and planting crops for two years before moving on to a new section. He planted sugar cane and coffee, built a *trapiche*, and began to raise cattle for market. By 1916, when Juan married María Vega Guzmán from El Queremal, he had a diversified farm operation. The sugar cane was processed into *panela* weekly. Another product was *aguardiente*, which was used to pay the workers. For household consumption there were maize, beans, cassava, plantains, yellow potatoes, *panela*, chicken, eggs, beef, pork, and milk. Milk was made into cheese, both for sale and for household consumption. Commercial products were brought up over the mountain to local markets in Bitaco and La Cumbre by horseback on trails and packed dirt roads.

Apparently this interior region between the main northern and southern transportation routes through the Dagua mountains had experienced considerable development since the end of the colonial period. The Álvarez family found some established farm communities, local market centers, and a network of roads and trails connecting farms and markets. This was the result of a steady process of settlement and development that had taken place in the mountain region between Cali and Buenaventura through the nineteenth century.

*Transportation Improvements.* The expansion and improvement of transportation routes in the region had played an important role in that development. These consisted of two types: routes that linked the Cauca Valley, and particularly Cali, to Buenaventura, to facilitate the external trade of the region; and routes whose main purpose was to facilitate intraregional trade. Most historical studies of the region have concentrated on the former. In the nineteenth century, the overland route between Cali and Buenaventura was improved, and a railroad line between those two towns was initiated.[15]

As we have seen, there had been sporadic and generally unsuccessful attempts to improve the Cali-Buenaventura route during the colonial period. These attempts intensified after independence. In 1829 Simón Bolívar created an organization to improve and administer the *camino de Anchicayá*, the route that followed the Anchicayá River inland and then passed through El Salado to Cali (Becerra Collazos 1977). During the next two decades there were many studies and proposals to build a hard-packed dirt road between Cali and Buenaventura, with those Cali residents owning land in the Raposo region pushing for the Anchicayá route. However, estimates of the Dagua and Anchicayá routes in the 1840s concluded that the Dagua route was preferable (Ortega 1920:453). Political power struggles in this period impeded the initiation of the project, and work on the road was not begun until 1845, with wealthy families from Cali providing the capital. After some disruption from the civil war of the late 1850s and early 1860s, and with support from a million-peso foreign loan taken out by the central government, the road was completed in 1866 along the Dagua River route (Cali-Vijes-Juntas-El Salto-El Saltito-Buenaventura, the northern route in map 2 (Gomez Benítez 1979).

15. Two detailed studies, from which much of the following discussion is drawn, are the thesis of Piedad Gomez Benítez on the *camino de Buenaventura* (1979) and the railroad history of Colombia by Ortega, vol. 1 (1920).

Because of lack of funds, however, the road was not well built. From Cordoba traffic still went by the Dagua River, with a trail for cattle transit (Ortega 1920:455). Pressure for transportation improvements between Cali and Buenaventura continued, particularly after the completion in 1855 of the Panama Railroad connecting the Pacific and Atlantic oceans. This gave the Colombian Pacific coast a new potential commercial importance, and interest in the Cali-Buenaventura route grew. The free-trade measures of the 1850s encouraged external trade and gave added economic incentive to externally oriented transportation routes. Weak central government support for internally oriented road construction and river navigation in the 1860s, limited by the fiscal resources of the state, turned into a fever of railroad construction in the 1870s fed by foreign financing. One of the routes to receive the greatest support was the Cali-Buenaventura line.

Alfredo Ortega (1920:21–38) details the protracted history of this project. Initiated in 1872, the line from Buenaventura was not completed as far as Cali until 1912. The original contractors laid only twenty usable kilometers of track by 1904, from Buenaventura to Cordoba. In 1905 new contractors reconstructed the line and extended it as far as Sucre before withdrawing. Finally, in July 1907, the Banco Central transferred the concession to the company that would ultimately finish the project. By 1909 the track reached to Papagalleros/Remedios, which was to become the town of Dagua. By 1912 the line crossed the mountains to Cali by way of La Cumbre. As can be seen from map 2, this route passed through the center of the *municipio* of Dagua, close to the settlements of La Virgen and El Palmar.

Although the purpose of the Cali-Buenaventura transportation routes was to serve the external trade of the Cauca Valley region, they also promoted the internal development of the area between Cali and Buenaventura. Settlers were attracted to lands near these routes, and they became means of linking the growing mountain settlements.

As map 2 illustrates, by the second half of the nineteenth century, there were three main external trails through the region: the northern Dagua River-Vijes-Cali route, the southern Anchicayá-El Salado-Cali route, and the *camino antiguo* (also referred to as the *camino real*), which cut through the middle of the region to connect the other two routes. By the 1870s, and possibly quite a bit earlier in some cases, major settlements along these routes included Juntas (later Cisneros), Papagalleros and Los Remedios (later joined as Dagua), Atuncela, El Salado, and El Carmen. A native of Atuncela, whose father was a muleteer in this period, described mule teams that would load cargo at Cordoba coming off the Dagua River canoes. Her father used the middle route, spending the first night in Atuncela, arriving at El Carmen after a second day's travel, and continuing to Cali the third day.[16] Products such as rice and salt were transported from Cordoba to Cali, while coffee was the major product being transported out by the late 1800s.

In the late 1800s and early 1900s, the construction of the Buenaventura railroad affected settlement patterns in this mountain region. Settlements flourished along the path of the finished track—Cordoba, Sucre, Cisneros, El Naranjo, Lobo Guerrero, Dagua, Lomitas, La Cumbre, Bitaco, and many others. Several of these were existing settlements given new life by the railroad. After 1888 Cordoba became an important terminal station for the railroad and was settled mainly by black railroad workers (Banderas 1944: 105).[17] The settlements of Papagalleros and Remedios eventually merged into the town of Dagua, which was officially founded in 1909 by the Pacific Railroad Company on land it donated. Dagua (for a brief time known as Caldas) became an important railroad terminal and the site of the railroad repair shop, where all the line's equipment was serviced. Because of the repair shop, Dagua soon attracted settlers from as far away as the northeastern mountains and north coast of Colombia (Becerra 1977: 7). La Cumbre, a settlement dating from 1852, was officially founded in 1913 by the railroad company when it built a station there (Banderas 1944: 192). The settlement of Juntas was renamed Cisneros after the railroad engineer, Francisco Cisneros, who had headed construction of the line from 1878 to 1886.

*Market Development.* The location of these settlements along the transportation routes suggests their market orientation, a hy-

16. There was some discussion between the informant and her daughters over whether the stopover was in El Carmen or El Salado. Although the informant was adamant that it was El Carmen, this remains uncorroborated.

17. Rail cargo was taken from Cordoba to Cali by mule until the railroad line

pothesis corroborated through the recollections of residents of the region. In addition to providing much of their own food needs, the settlers geared their production to a variety of markets: the food needs of the mining population, of the town of Cali, and of the workers constructing the transportation routes, the exchange of products between settlements, and, ultimately, production for export markets.

As early as the eighteenth century, some settlers, as we have seen, may have supplied food to the Raposo mining population. This was probably particularly the case with the settlements in the Anchicayá region, although there was some mining activity as far north as Dagua. The decline of slavery over the first half of the nineteenth century dealt a heavy blow to this mining region, however, and the abolition of slavery in 1851 effectively ended its economic importance.

However, commerce, and with it the town of Cali, was growing in the nineteenth century. The average annual rate of growth of the population of Cali between 1806 and 1906 may have been as high as 3.3 percent.[18] By the early 1800s there were signs that, aside from meat, food scarcity was becoming a problem in the city. A measure was passed requiring the haciendas to cede one-third of their land to the city, to be distributed to its landless residents. One of the goals of this redistribution was to help resolve the growing problem of food scarcity (Pacheco 1980:20–24).[19] The city's food was being provided by the smaller haciendas, as well as by the valley's growing group of medium and small holdings. Located closer to the city than the *haciendas mayores*, the small haciendas marketed plantains, maize, cassava, *miel*, and *panela*, as well as meat (Escorcia 1980:56). However, food production in the valley must

was completed in the early 1900s. There was also a trail to the Calima River in the southern Chocó.

18. This rate is based on the population figures presented by Camacho Peréa (1962:94). According to the figures presented by Diaz del Castillo (1937:100), the rate would have been 2.3 percent. However, Diaz's figure for 1928 is wrong, which casts some doubt on his other data.

19. These land concessions were called *ejidos*. Another goal of *ejido* policy, according to Pacheco, was to provide a new source of city revenue. Land redistribution would also reduce the social pressures bound to result from a situation of an increasing population of mestizos, poor whites, and free slaves in an area of highly monopolized landholdings. However, as Pacheco points out, the hacienda

have been limited by the amount of valley land held by the latifundios and the large haciendas, whose only locally marketed food product was beef.[20] As a result, by 1849 the newspaper *La Opinión* was discussing the "high cost of living and the scarcity of foodstuffs" in Cali (Pacheco 1980:24).

As mining activity decreased and the demand for foodstuffs in Cali grew, it appears that the region of El Salado emerged as an important nucleus of smallholding production to meet urban food needs. By the late 1800s, a dense network of trails crisscrossed this area and linked it to Cali. El Carmen had become a central market site for the region, with a network of trails emanating from it as can be seen from map 2: to Cali by way of San Bernardo, Felidia, El Saladito; to El Queremal; to Dagua by way of Platanales and Calaveras (and then on to Buenaventura by mule). El Carmen was also linked via El Salado to the Anchicayá route, and to Dagua through another trail by way of El Limonar. The farmers produced maize, kidney beans, cassava, *arracacha*, sugar-cane products (*miel*, *panela*, *aguardiente*), and dairy products (milk, cheese), and also raised cattle and chickens. The most important marketed items were the sugar-cane and dairy products, livestock, and beans—high-value products in relation to their bulk and weight, a key consideration given the difficulties of transporting products to market.

The railroads affected agricultural production as well. According to settlers of this period, farmers in the area grew foodstuffs for the construction and repair workers. By the early 1900s, sugar-cane products were important cash crops for Juan Álvarez and his neighbors. The town of Dagua became a thriving market center because of the railroad activity and its growing population. In 1918 it was renamed the government seat of the *municipio* of Dagua.

In addition to production for regional and urban markets, and sporadic production for the transportation workers, the settlers of

owners resisted the ceding of *ejido* lands. This became an important source of conflict in the city.

20. According to Escorcia (1980:56), "there is no evidence that the haciendas in the nineteenth century increased their agricultural production to satisfy the demand of the population in the urban centers like Cali." There is evidence, on the other hand, that the haciendas were increasing their production of tobacco and quinine for export at the expense of domestic foodstuff production (Hyland 1982:385).

the Dagua region were supplying another market by the late 1800s: the international coffee market. As early as the 1880s, settlers in the mountains around Atuncela and El Naranjo were planting coffee trees.[21] Coffee cultivation appears to have become widespread in appropriate climatic zones throughout the Dagua mountains by the early 1900s, although it did not eclipse other production.

*Access to Land.* Transportation improvements and market growth throughout the nineteenth century provided an increasingly favorable environment for potential settlers. Another key condition needed by these settlers was access to land. The history of the Álvarez family shows that *baldío* land was still available in this region in the late 1800s. In general, early settlers in the El Palmar area were able to claim *baldíos*.[22] By 1910 most of the *baldíos* in the area had been claimed, and new arrivals had to buy land from those who had gotten there first. Transactions involved purchases of improvements made to the land—*compras de mejoras*—rather than purchases of the land itself, since the settlers had not obtained formal title to the land. Many migrants arrived without funds to buy *mejoras* and started out as agricultural laborers. This was the case with ten of the thirteen early migrants interviewed. Another worked for the railroad. Only two of the thirteen acquired land immediately upon arrival, and both arrived before 1920. In this period, however, it was not difficult to accumulate sufficient funds to buy *mejoras* in a very short period of time. All thirteen men soon acquired land in El Palmar.[23] A few examples will serve to illustrate this process.

Early arrivals like Juan Álvarez found it relatively easy to establish themselves in the Dagua mountains. Juan continued to acquire land, both in La Virgen and elsewhere. In the 1920s he bought land in the region of El Salado, which he sold in the early 1930s to buy land in Digua: pasture land, "steep and full of snakes," to the

21. The source for this information was an interview with a former Atuncela resident.

22. This section is based on interviews with thirteen male heads of household who had arrived in El Palmar before 1940.

23. This does not necessarily mean that all migrants arriving in this period acquired land in El Palmar, since those who did not might have soon moved on. We would have no record of this, only of the "success" stories. If there were early migrants who failed to acquire land, none of them remained in El Palmar.

west of El Queremal (see map 2). By the 1940s, Juan had acquired well over 150 acres of land in La Virgen. Later arrivals, too, were prospering. Manuel Gratiniano Valencia arrived with his family in 1918 from Rosario, Cauca, where he had sold his small farm.[24] He used the proceeds to buy twenty acres of land in El Palmar. By 1920 he was able to purchase twenty-five more acres of land, and in 1923 another thirty acres, for a total of seventy-five. He had a diversified farm in food crops, sugar cane, and coffee.

In 1923 Aristides Benavides, then sixteen years old, arrived in El Palmar and began to work for Manuel. Aristides had again been bitten by the urge to wander and had left home, this time for good, at the age of thirteen. He had come to Valle and worked at a number of jobs for several years: as a wage laborer on "La Margarita," a *hacienda menor* supplying Cali with beef, wheat, maize, and onions; in the *galería* in Cali, a market where produce from the surrounding area was sold by vendors to urban consumers; and as a wage laborer in Felidia (see map 2), an agricultural settlement with many family farms growing crops for Cali. In 1921 he had left Felidia for El Queremal, where much of the land was still *baldío*. Aristides had claimed five hundred hectares, but could only cultivate fifteen acres. His land had been at a fairly high altitude, too cold for coffee and many food crops, but good for cattle, which were sold in the nearby market town of El Carmen. In 1923 he had sold his *mejoras* for fifty pesos, bought four horses, and come to El Palmar.

El Palmar and the neighboring community of La Virgen were already well populated. There were some twenty families in El Palmar living in the dispersed settlement pattern common to Colombia's Andean peasant communities (Fals Borda 1955), and all the *baldíos* had already been claimed. After a year in El Palmar, Aristides married Manuel's daughter Romelia. With his earnings from agricultural labor and the thirty-three pesos he got from the sale of three horses, he bought nine acres of land, planted in coffee, plantains, and sugar cane, from his father-in-law. Aristides gradually built up his holdings. In 1931 he bought six acres of improved land in El Queremal, in coffee and sugar cane. He then claimed five hundred acres of *baldíos* in the mountains of El Queremal,

24. Based on an interview with Manuel's son, Miguel Valencia, in 1979.

which he began to clear for livestock. He bought and sold some land in Digua. He also bought more land in El Palmar: five acres, three of them from a brother-in-law. In 1933 his wife acquired ten acres as her inheritance from her parents, and this land, too, came under Aristides' management.[25]

Certain aspects of this process of land acquisition in El Palmar are noteworthy in light of the existing literature on this topic. First, there were no formal land claims made in this period—no official registration of landholdings. Rather, the farms in the area were marked out with trenches or with fencerows of *pinuela* (a sharp-spined bush). This means that formal records of *baldío* distribution, frequently relied on to give a picture of the settlement process in this period, would contain no evidence of this smallholding settlement. Much of the nineteenth-century settlement of the Colombian mountains by mestizo farmers may have followed this pattern, leaving no trace in the official records.[26] Despite the lack of formal title, land claims were respected by later arrivals to El Palmar.

Second, there was an active market in land improvements even in the earliest stages of settlement. There was also an active wage-labor market. Both of these reflect the market orientation of the early farmers. The economy that they were creating in the mountains was a monetized one, which reacted to changing product and factor market conditions. Although directly producing much of the food needs of their households, these farms were far from the isolated, subsistence peasant holdings frequently suggested by the literature.[27]

Third, the settlement of this area appears to have been free of

25. Although in theory partible inheritance gave women direct access to land, in practice during this period (and frequently today as well) land was turned over to a male (husband, son, or brother) to operate.

26. Catherine LeGrand's detailed study of public land disposition between 1870 and 1936 suggests that this was the general pattern (LeGrand 1980; see especially pp. 183–85).

27. LeGrand (1980) finds this active market orientation to have generally been the case in settler communities throughout Colombia. Chapter 4 of her work provides a detailed discussion of this process. This contrasts with the statement by López that the smallholders were noncommercial farmers: "This subsistence production [in both the eastern and western mountains] lacked a commercial character (there was no production for the market)" (López 1975:51). This is a commonly held view that recent research is beginning to prove erroneous.

the conflict between large landholders (or pretenders to that status) and mestizo settlers that characterized the colonization process in some other parts of the country.[28] Although there were, as we have seen, extensive haciendas in the Dagua region during the colonial period, much of the land was still in the national domain. The haciendas themselves underwent a process of fragmentation in the 1700s and 1800s,[29] and it is likely that the severe decline of mining activity, the overall labor scarcity, and the poor transportation development of the region discouraged the recomposition of large estates in this area during the nineteenth century. To this day, landholdings in Dagua are less concentrated than for the country as a whole: in 1970, 55.6 percent of Dagua's farmland was in holdings of over 250 acres, compared to 67.5 percent nationally (DANE 1973).

*Labor.* As a final condition, the settlers needed labor to work their expanding holdings. As a single young man in the sparsely populated mountains of El Queremal, Aristides Benavides had been able to work only a small portion of his claim; this was one factor causing him to leave El Queremal for El Palmar. There he was able not only to acquire land, but also to marry and begin to build his family labor force.

More established family-farm communities, such as Felidia, already contained well-developed wage-labor markets on which the farmers could rely to supplement family labor. In El Palmar, too, wage laborers were important to farmers such as Manuel and Juan who were building large farms while their children were still young. The wage-labor market in this frontier community was not well developed, however, and the farmers could not always count on finding workers when they needed them. The neighbors sometimes worked together in *mingas* for the big tasks such as clearing plots, harvesting, or house raising, but this, too, failed to provide a steady

28. This conflict has been well documented. See, for example, chapter 6 of LeGrand's work (1980), in which she shows how the settlers (*colonos*) without land title were dispossessed by the large landowners, often many years after the settlers had established themselves.

29. Colmenares (1976:251–53) provides much information on the breakup of the Papagalleros estate, for example.

and dependable supply of labor for day-to-day farm operations. For these, the farmers hoped to be able to draw on their families.

Families were established quickly and, as is still the case today in El Palmar, children participated in farm and household chores at a very early age. Young children helped with such tasks as bringing water from the stream to the house, washing maize, carrying plantains, gathering wood, bringing root crops from the garden, aiding in the harvest, aiding with the processing of the coffee or the milling of the sugar cane, and even hauling cane *a puro hombro* (on their shoulders). As Aristides Benavides described it, his children "weren't allowed to lose a second." Where resources were scarce, patriarchal norms allowed children's consumption to be kept to a bare minimum. Juan's children, for example, had no shoes and wore the same clothes all week. The family lived for several decades in shacks that he built. The growth of his family did not prevent the patriarch from directing most of the family's earnings into the expansion of his farm.

This reliance on family labor, and the patriarch's ability to maintain family consumption at low levels in order to channel resources into farm development, created strong incentives for him to desire large family size.[30] The preference for high fertility was compounded by the "flow" of sons out of the patriarchal unit in response to the internal tensions discussed above. A similar process occurred with the daughters, who left most frequently to take their places in new patriarchal units. The separation of children from the family unit as they matured resulted in a continual attrition of the labor force

30. In general, the patriarch would desire more children as long as the addition to family sustenance provided by the labor of the new member was expected to be greater than the consumption needs of that individual. There is a growing body of literature attempting to measure the value of children to parents in rural areas of developing countries. While Mueller (1976) found that children were a net cost to their parents, Nag (1972), Nag, Peet and White (1977), and Caldwell (1982) found that children were a net benefit. Caldwell (1982) has concluded that in traditional households, the expected flow of income is always from offspring to parent and that maximum fertility is therefore desired by the parents. These analyses do not distinguish between the benefits expected by the patriarch and those anticipated by his wife, and therefore fail to consider the possibility of divergent fertility preferences on the part of the patriarch and his spouse. This issue is considered in more detail in chapter 6, where I discuss changing fertility trends in El Palmar.

upon which the farm depended. The solution was to bear children at regular intervals for as long as possible, so that there would always be younger offspring to take the place of those who were leaving.[31]

The result of these factors was a tendency towards quite large completed family size. For those ten women in El Palmar who were seventy years of age or older when interviewed in 1978–79, average completed family size was 6.2 children, counting only children still alive at the time. Given fairly high rates of infant and childhood mortality, and the number of offspring who might have died at a later age, this figure suggests that these women were regularly pregnant throughout their childbearing years. Indeed, one woman with ten surviving children in 1978 had actually given birth to eighteen children. Both Romelia Benavides and Maria Álvarez bore children regularly. Romelia had nine surviving children, while Maria had ten surviving children and two who died in infancy.[32]

## The Political Economy of Family-Farm Colonization in Colombia

In general, despite sometimes intense conflict between settlers and large landholders, a comparable process of mestizo settlement was taking place throughout the mountains of Colombia in the nineteenth and early twentieth centuries. Most existing research has focused on the appropriation of national land for export production in this period—particularly tobacco in Ambalema and coffee in Cundinamarca, Antioquia, and later Tolima.[33] Not surprisingly,

31. The replacement of female labor was as important as the replacement of male labor, although this is often overlooked. It was clear from the examples of farm families in El Palmar that the chores delegated to the wife (food preparation, child care, laundry, house cleaning, bringing food to the fields, and numerous farm chores, including care and feeding of small animals, care of garden plots, processing of coffee, and harvest labor for a variety of crops) could not be completed without the assistance of young children and particularly adolescent daughters.

32. Rapid population growth was a notable characteristic of Colombian family-farm communities as early as the nineteenth century. The most outstanding examples are the Antioquia communities (see Parsons 1968: chap. 8).

33. In addition to LeGrand 1980, studies of the appropriation of land for the establishment of export estates can be found in Safford 1966 and Palacios 1980. Discussion of the acquisition of land in Cundinamarca and Tolima as large cattle,

given the high potential value of export production, these chronicles find much evidence of conflict in this process. Yet even in these regions we find the development of smallholding communities, with Antioquia being the most notable example.[34] Why was this allowed—even encouraged—by successive nineteenth-century governments that are generally understood to have represented elite interests?

One possible explanation is the support that such smallholding communities could give to commercial expansion. We have seen the development of economic activity in Dagua as the smallholding population expanded. The benefit to commercial interests would be even greater in those areas, such as Antioquia, where coffee cultivation could support a more active trade. The growing wealth and power of the Antioquia coffee merchants is well documented. The small coffee farms also gave rise to the agricultural equipment industry, particularly the production of small depulping machines. Not surprisingly, the commercial bourgeoisie of the cities of Antioquia were strong supporters of the colonization movement (Parsons 1968:76–77; Ospina Vásquez 1955:311; Palacios 1980:162; Fals Borda 1969:98; Villegas 1977:9–10).

This potential for commercial development may explain in part the developing ideological commitment within certain sectors of the elite to the expansion of smallholdings, particularly after 1850. This was reflected in legislation progressively more favorable to colonization over the course of the century (Fals Borda 1969:98; Hirschman 1963:100; LeGrand 1980:53–74). Nevertheless, the image of Liberal commercial bourgeoisie pitted against Conservative landed elite would be a highly simplified representation of a complex

---

sugar, cinchona, tobacco, and coffee estates can be found in Mejía 1971:134, 146–47, 157, 169; Ospina Vásquez 1955:242; Hirschman 1963:98–100; Safford 1966:221–22, 280–81, 297–300; Fals Borda 1969:100; Parsons 1968:137–41; and Villegas 1977:57–61.

34. The most complete study of the Antioquia smallholding settlement process is Parsons 1968, especially chaps. 5 and 6. Villegas (1977:68–70) presents a detailed discussion of the settlement process in Antioquia, Cundinamarca, and the Santanders and Boyacá that shows the varied outcomes of the struggles of the settlers with the large estate owners. Settlers appeared to have greatest success in the Santanders and Antioquia, regions where they were mainly "whites" ("*blancos*"—actually predominantly mestizos) who could not be as easily enserfed as the indigenous population of Cundinamarca and Boyacá.

situation.[35] In fact, the agreements that were reached often benefited the preexisting latifundios in the regions of settlement. In Antioquia, as in the mountains of Dagua, latifundio land had been virtually worthless: poor soil, no labor force to work it, in unpopulated, geographically isolated parts of the country. In the agreements with the settlers in Antioquia, the estate owners were frequently indemnified for land adjudicated to the settlers. The value of the rest of their land was then considerably increased by the settlements and by the transportation improvements and trade that followed (Parsons 1968:chap. 6; Villegas 1977:13–16).

Even in areas where commercial capital was being channeled into the creation of export estates, as in Ambalema in the 1850s and Cundinamarca in the 1890s (Safford 1966:1–3, 201–2, 211–21, 240; Palacios 1980:37–47), the relationship between large estate and smallholder was not necessarily one of conflict. While struggles could and did develop over land, the incipient estates in these largely unpopulated frontier regions had one crucial requirement: a labor force. Although labor was brought to these estates from the highland areas by labor contractors (*contratistas*), the permanent migration of peasant families from the highlands to the frontier regions played an important role in the provision of a stable local labor force.[36] Thus again, as with latifundios in Antioquia, a controlled process of settlement could benefit the export haciendas, although the potential for conflict over land was ever present. In other regions, as the experience of Aristides' family in Tambo illustrates, potential settlers could provide an important temporary labor force for the commercial operations of food-producing haciendas.[37]

35. Colombian politics have been dominated by two parties, the Liberal and the Conservative, since the middle of the nineteenth century. LeGrand (1980:26) argues that changes in land policy reflected the opinions of members of both political parties.

36. According to Palacios (1980:68–71), the labor force brought by the *contratistas* to the Cundinamarca coffee plantations was mainly the seasonal labor force needed for the most labor-intensive tasks. Families from the fragmented smallholding zones in Boyacá also migrated permanently to the new coffee plantations as a labor force. Villegas (1977:63–65), drawing on the nineteenth-century writings of Medardo Rivas (*Los trabajadores de Tierra Caliente*), discusses the role of this migration in the development of the Ambalema estates.

37. Villegas (1977:58) describes the use of immigrant labor on the expanding

Another aspect of mestizo colonization that some members of the elite may have considered was the effect of this process on the regions of origin of the migrants. For the estate owners in these areas, the emigration of mestizos to frontier regions could be viewed as either a positive or a negative phenomenon. On the positive side, it could reduce conflicts with land-hungry peasants pressing up against the obstacle of latifundio land monopolies. Such conflicts were intensifying in the nineteenth century throughout the more densely populated regions (Zuleta 1973:41). On the other hand, a too-rapid process of emigration could lead to labor shortages for these latifundios. Support for a policy of public land distribution to mestizo settlers would thus vary, depending on whether the estate owners viewed the major problem of the moment as conflict over land or labor scarcity.

Colonization was therefore a contradictory process, not only because it could have different effects for different groups, but also because it could have different consequences for any particular group at different times. As a result, while public land legislation after 1850 generally professed to be pro-colonization, the actual implementation of public land policy varied markedly (LeGrand 1980). In Antioquia, for example, one can find examples of land conflicts resolved in favor of the settlers, and examples of those resolved in favor of the haciendas (Villegas 1977:10–54). The same is true in other frontier areas. Overall, there was enough disagreement over land policy to provide an opening for massive population movements and smallholder colonization, particularly in areas outside the principal centers of export agriculture development. Even in those areas, there were enough members of the elite who might benefit from this process that it was sometimes permitted or even encouraged.

This is not to suggest that the settlement process went smoothly in nineteenth-century Colombia, or that the smallholders received the unequivocal support of national and regional governments. The period was rife with land struggles in the regions of settlement and labor struggles in the regions of origin. Most public land was being

cattle estates of Cundinamarca, with families coming from as far away as Antioquia to settle in the area.

acquired by the large estates (Palacios 1978: 175–78; Villegas 1977; LeGrand 1980). Yet considerations operated in favor of the settlers that should not be overlooked, and neither should the extent of the colonization movement be understated. This movement gave rise to a nucleus of family-farm communities in the mountains of Colombia, settlements, such as those in Dagua, that played an important role in regional and national economic life. The early history of El Palmar has allowed us to examine this settlement process in greater detail than has previously been possible.

*Chapter Three*

# The Changing Context for Family-Farm Production in the Early Twentieth Century

In chapter 2 we saw how the necessary conditions for the prosperous development of the family-farm economy—access to land, labor, and markets—were met in the mountain communities created during the migratory movements of the nineteenth and early twentieth centuries. In this early period, farmers in these settlements had a considerable advantage over those in the more densely populated areas of colonial settlement. The continued success of these settlers as farm households was not assured, however. It depended on continued access to land, labor, and markets, conditions that would be affected both by developments "external" to the family-farm economy (overall patterns of economic development, market development, transportation improvements, and the development of the estate sector) and by developments "internal" to it (relations between family members that conditioned demographic trends, family labor-force development, and intergenerational access to land).

The family farm was affected by changes in both sets of factors during the twentieth century. In this chapter I consider the impact of the far-reaching transformation of the Colombian economy that began to take place in the 1920s and 1930s. I shall show that the complex market development of this period was favorable to the patriarchal farms, and examine the reasons for the continued ability of these farmers to compete with the large agricultural estates.

## The 1920s: Structural Change and the Expansion of Family Farming

In the period of settlement of the Dagua mountains, the farmers could count on an abundant supply of virgin land and on a growing family labor force to develop their holdings. Yet in the nation as a whole, these factors were becoming less favorable. By the 1920s, the mountain frontier in Colombia was closing. In regions such as Dagua, among the last-settled mountain zones, small pockets of *baldíos* remained (for example, in El Queremal and Digua), but these tended to be the most marginal land. Large expanses of unclaimed land existed only in lower-lying areas of the country, such as the plains (*llanos*) of the southeast and the northern coast. Parts of these areas had been incorporated into the estate sector for cattle raising in the 1910s and 1920s (Ospina Vásquez 1955:354). In the rest, problems with climate, geographical isolation, and particularly disease effectively prevented settlement at this time.

While the nineteenth-century outlet for population growth in the countryside was ending, that growth was continuing, fed by the large families of patriarchal farm households in the mountain ranges. Between 1905 and 1938, the population of Colombia increased from 4,488,782 to 8,702,000, for an average annual growth rate of 2.8 percent (see Appendix A). With colonization a less viable option, conflict between the mestizo farmers and the large landowners threatened to intensify, both in areas of traditional settlement and in regions where the commercial hacienda predominated.

Although access to land as a means of expanding the family-farm economy was becoming more limited, other factors were still favorable. The tension inherent in the closing of the mountain frontier was moderated by major changes taking place in the Colombian economy. Coffee demand in the United States and Europe increased in the early 1900s, and Colombia rapidly became one of the world's major coffee-producing nations.[1] The rapid increase

1. Most general histories of Colombia document the rise of coffee production in the late 1800s/early 1900s. In addition, there are several comprehensive studies of Colombian coffee cultivation, the most recent being Palacios 1980. See also Parsons 1968; Machado 1977; and Bergquist's chapter on Colombia (1986). The most rapid growth in coffee production occurred after World War I. According to Pala-

in Colombian coffee exports, from 548,000 60-kilogram sacks in 1910 to 3,118,000 in 1930 (Mejía 1971:222; Palacios 1980:232), spurred fundamental structural changes in the Colombian economy. The growth of trade and commerce centered around the coffee economy was accompanied by urban expansion, massive spending on public works, and industrial development. Imports grew from an annual average of $19,346,000 in 1901–13 to $94,927,000 in 1921–29 (Fajardo 1983:39), while Colombian industrial capacity increased over 50 percent during that same period (Mejía 1971:245).[2]

The most dynamic coffee-growing activity had shifted by this time from the eastern mountains of the Santanders and Cundinamarca to the central and western mountains of Tolima, Antioquia, Caldas, and Valle.[3] This expansion of coffee cultivation in the western mountains of Colombia in the early 1900s, and particularly the 1920s,

---

cios (1980:198), coffee production grew at an average annual rate of 9.9 percent between 1915–17 and 1930–32.

2. Much of the industrial investment was Colombian capital, particularly Antioquia merchant capital seeking new outlets after the opening of the Cali-Buenaventura railroad cut into its control of the commerce of the Southwest (Mejía 1971:245; Bejarano 1975). This investment responded to the growing internal market. In Antioquia itself, small and medium coffee producers had been an important market for hand-operated coffee depulpers and other agricultural equipment since the late 1800s (Ospina Vásquez 1955:311; Palacios 1980:152–53). By the post–World War I period not only the coffee growers but all the other sectors linked to coffee began to offer a market for industrial products: the smallholding producers of fiber (*fique*) coffee sacks, the work force for the transportation improvements, and the workers and artisans of the expanding commercial, trade, and shipping centers of the west and of the Caribbean coast were all acquiring purchasing power (Londoño and Velasquez 1974:117–18; Mejía 1971:223–24; Bejarano 1975:245–47). McGreevey (1971:236) argues that the equitable distribution of income in the coffee economy generated demand for cheap textiles, stimulating the early development of the cotton textile industry in Medellín. By cutting off traditional imports, particularly textiles, from England, World War I had given Colombian industrial development its start (Ospina Vásquez 1955: 312–13, 386–88). But the real spurt in industrial development occurred in the 1920s, when industrial investment was encouraged not only by the expanding internal market but also by a flood of foreign capital. Mejía (1971:244–45) argues that this capital, while it did not go directly into industry, lowered interest rates and pushed domestic capital out of speculative and into industrial investment. Much of the foreign capital inflow financed unprecedented government expenditures on infrastructural development, both transportation improvements and urban public works, that was the essential underpinning to industrial development (Bejarano 1975:298–99).

3. By 1932 these four regions produced 70 percent of Colombian coffee, up

provided a double incentive for transportation improvements in the region. First, an increasing volume of trade was passing from the western departments through Cali and Buenaventura: coffee on the way out, a variety of imports on the way in. The opening of the Panama Canal after World War I decreased transportation costs through Buenaventura and stimulated trade through this port. Second, government revenue from trade taxes provided funds for the expansion of public works throughout the country. These funds were supplemented by massive loans from foreign sources, particularly the United States, to finance infrastructural development. New long-term loans from U.S. sources to various levels of the Colombian government amounted to $171,000,000 between 1920 and 1928 (Rippy 1976:154–55).

By 1928 the Pacific Railroad had completed lines linking Caldas and Quindío (coffee-growing regions of Antioqueño settlement) with Cali and Buenaventura. Coffee was shipped from Buenaventura to San Francisco and through the Panama Canal to eastern U.S. ports (Palacios 1980:153; Parsons 1968:147). Transportation improvements in this period included highway as well as railroad construction. Plans were finally laid in the early 1920s for a new Cali-Buenaventura highway. Work on the road was initiated from Cali in 1926 (Banderas 1944:78), proceeding along the middle route through the Dagua mountains (see map 3): through El Saladito, to a settlement that came to be known as "Km 18," passing north of El Carmen through the settlement of Olaya Herrera (or El Treinta, kilometer 30), and on to Dagua. Although construction of the highway was halted in the late 1920s—there was a continuing dispute over the location, with some influential figures pressing for the Anchicayá route—the completed link from Dagua to Cali improved access to both markets for the farmers of the Dagua mountains. The older residents of El Palmar still remember the inauguration of this spur of the highway in the late 1920s. From Dagua, traffic went to Buenaventura by railroad or by mule trail.

The combination of rural population growth, the closing agricultural frontier, and the growth of employment opportunities in transportation improvements, urban public works,[4] trade, and in-

---

from 3.6 percent in 1874 and 45.5 percent in 1913 (calculated from McGreevey 1970:210, table 1).

4. In Cali, for example, the 1920s saw major urban development. According to

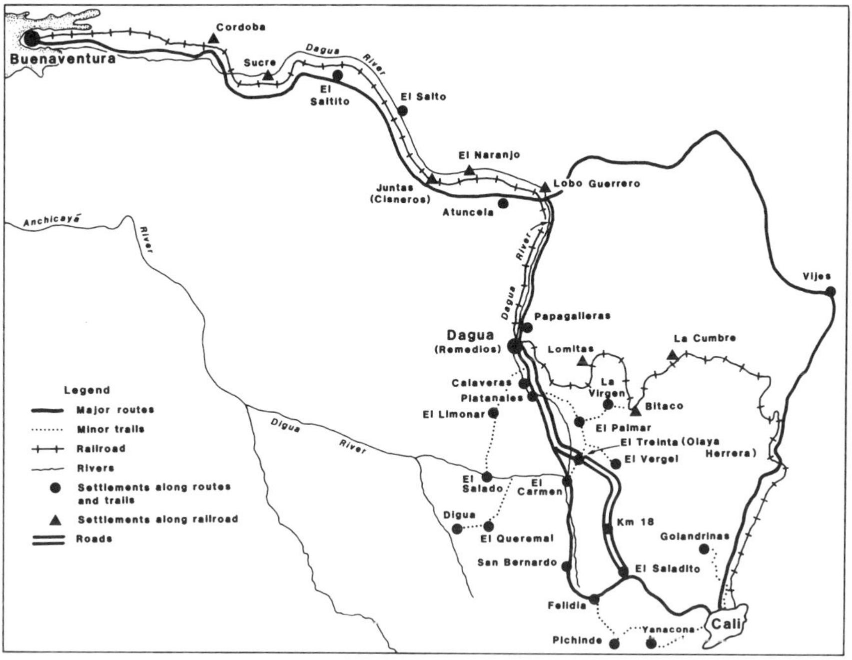


Map 3. The Road to Dagua, 1926.

dustry led to large-scale migration to the cities in this period.[5] The rapid growth of Cali is representative of this urban migration: between 1918 and 1928, the population of the city grew from 45,600 to 59,500 (Camacho Peréa 1962:94), an average annual growth rate of 3 percent.[6] The expanding wage-labor force in public works and

---

Diaz del Castillo (1937:130), the "progress of Cali" was initiated in this period. Before 1920 the streets of the city, paved by the time of his writing in 1937, were "almost as they were in the colonial era."

5. For an overview of urban population growth in this period, see López 1975:73–76. According to his figures, urban growth rates in cities of over 1,000 inhabitants averaged 4.7 percent between 1918 and 1930.

6. Diaz del Castillo (1937) gave a figure for Cali of 110,009 in 1928, but this seems much too high given subsequent figures, such as that of 88,366 recorded in the population census of 1938 (DANE 1942). It is likely that the figure of 110,009 was for the *municipio*, rather than the city, of Cali. Although some industry was being established in Cali, its growth during this period was based mainly on the expansion of trade and commerce in the region. Ospina Vásquez (1955:391–98) contrasts the industrial development of Medellín with the virtual absence of such development in Valle in the 1920s.

industry, along with the growth of trade and commerce, led to an increasing monetization of the economy and increased demand for marketed foodstuffs.

The impact of these changes on the agrarian sector was complex. Overall, I would argue that the economic development of the period fortified the peasant economy, increasing the competitiveness of this sector with the estate sector and improving the ability of family farmers to earn a livelihood from their farms. The increasing demand for marketed foodstuffs drove up agricultural prices. Despite a sharp reduction in tariffs on agricultural imports legislated in 1926, which had its greatest impact on grain prices, peasant-produced food prices in Colombia's fifteen major cities were still markedly higher in 1929 than they had been in 1923 (López C. 1975:89). With most inputs to peasant production available on the farm (seed, compost, handmade tools, family labor), rising agricultural prices enabled a given amount of farm land to support a larger farm population.

We can clearly see the impact of these developments in the mountains of Dagua. The need to provision the workers on the Cali-Buenaventura highway, whose numbers grew to 525 (Banderas 1944:78), and the growing demand of the rapidly expanding population of Cali, stimulated agricultural production. Aristides Benavides' account of his wanderings in the vicinity of Cali in the early 1920s provides a rich chronicle of the forms that production took, from *haciendas menores* such as the one on which he first worked, to the small and medium farms of Felidia, El Saladito, and other mountain communities near the city. Further from Cali, there were important mountain markets by the early 1920s in El Carmen and La Cumbre on Saturdays, and in Dagua and Bitaco on Sundays. A network of trails linked these market towns to the farm settlements, as can be seen from map 3.

Farmers in the Digua/El Queremal/El Salado region traded foodstuffs in El Carmen, as did the farmers in El Vergel. Those in the El Palmar/La Virgen area had a choice of markets. Some brought their products to Bitaco and La Cumbre. Others marketed in Dagua, home by the 1920s to over five hundred railroad workers and their families. The trails to Dagua were difficult before the construction of the new Cali-Dagua road, however, and many of the farmers in this area preferred to go to the Saturday markets in El Carmen, to which purchasers regularly came from Cali.

The shifting of coffee cultivation to the western mountain ranges of Colombia was accompanied by the growing participation of peasant producers in that cultivation. By the 1920s, as the stories of Juan and Aristides suggest, coffee was becoming an important crop in the area of El Palmar, although by no means dominant. The extent to which farmers in El Palmar were cultivating coffee in the 1920s is indicated by the fact that of the parcels acquired in the 1930s by households interviewed, 60 percent had coffee.[7] According to some of these farmers, they switched crops depending on relative prices: "Buscamos lo caro" (we look for high prices).

The participation of the Dagua smallholders in the expanding coffee and foodstuff markets of the 1920s was not an isolated phenomenon. Machado's recent analysis of the developing coffee economy presents a clear picture of the expansion of coffee cultivation on peasant holdings (Machado 1977: chap. 3; Palacios 1980: 228–29). In 1932 the first national coffee census found that 87 percent of Colombian coffee farms had fewer than 5,000 trees (i.e., under four acres in coffee). In Valle del Cauca, the figure was 92 percent (Bejarano 1979: 53).

According to the classical Marxist analysis of agrarian development, reviewed in chapter 1, we would expect the kind of market expansion taking place in Colombia to lead gradually, but inevitably, to the displacement of the peasant producer. We see instead the expansion of peasant production in response to the new market opportunities of the 1920s. Peasant producers were competing successfully in domestic markets with food producing haciendas and in international markets with the Colombian coffee estates. To what was this continued competitiveness due?

There is evidence that the conditions of the 1920s raised the labor costs of the estate sector. Hacienda owners (*hacendados*) in that period complained bitterly of the public works expenditures and urban development that drew workers out of the agricultural labor force (Salazar 1973: 45–46).[8] This was particularly the case with the coffee haciendas of Cundinamarca and Tolima. Histori-

7. Based on the 1978–79 interviews with 105 resident households.

8. Fajardo (1983: 36) and Machado (1977: 298–99, 302–30) describe proposals by the landed elite in the 1920s to restrain labor mobility and to import labor. These proposals generated heated debate, with a number of writers pointing out that the problem was not an actual *scarcity* of labor, but rather a shortage of labor at the level of remuneration that *hacendados* had been accustomed to paying.

cally short of labor during the peak harvest period, these haciendas were threatened by the migration of agricultural workers to the city. The food-producing haciendas were also affected by the developing labor scarcity and rising wages. As with the coffee estates, these haciendas relied on a combination of wage labor and usufruct arrangements with tenant workers, examples of which we have seen with the experiences of the Cabreras and the Benavideses.

## A Theoretical Perspective on the Persistence of Peasant Production

A number of writers have argued that these labor difficulties on the estates improved the competitive position of the family farms. In asserting that these farms were less susceptible than the estates to the changing labor conditions of the period, these writers implicitly or explicitly rely on what I termed in chapter 1 the "behavioral advantage" theory of the family farm (Machado 1977:127–28; Fajardo 1983:26). By drawing on family labor at an implicit wage below the market wage, the family farm could insulate itself from rising labor costs and compete with otherwise more efficient larger farms.

This is not, however, a sufficient explanation of the competitiveness of the family farm. We have seen that these farms, while relying primarily on family labor, also depended on hired labor. This was particularly the case for the coffee harvest, but was also true of many farm operations in the case of young households with little family labor to draw on. Furthermore, might not the employment opportunities of the period be expected to have drawn family labor away from the patriarchal unit, and increasingly to have forced the patriarch to consider the opportunity cost of that labor? True, the wage-labor market may not yet have been sufficiently developed, or the authority of the patriarch over family labor may have been sufficiently strong to withstand this pressure, but the ability of the family farm to keep labor costs down cannot automatically be assumed.

It is true that, as the behavioral cost advantage theory postulates, the independent family farms of this period did not require prices sufficient to generate rent or profit, which may have given them some advantage over the larger estates. But would we not expect that the greater productivity of the large farms could outweigh this advantage, as well as any labor cost advantage the smaller farms might enjoy?

In order to understand the continued competitiveness of the family farm, a deeper analysis of the nature of agricultural production is needed. In its traditional or natural state—uninfluenced by modern scientific and technological advances—agricultural production is a biological, sequential, spatial, and seasonal process. With some exceptions, the tasks must follow a sequential pattern (soil preparation, planting, weeding, harvesting) with a specific maturation period, determined by biological processes, and therefore a seasonal pattern of labor requirements. These distinctive features of agricultural production may give small-scale, family-labor-based production a *technical* cost advantage over large-scale units, thereby posing obstacles to the development of capitalist agriculture. Here we can find the underlying explanation for the persistence of family farm production in the early stages of capitalist development in Colombia.

First, the biological and therefore sequential nature of most agricultural activities limits the possibilities of task specialization, one of the basic sources of economies of scale in production. One worker cannot always be plowing, another harvesting, and so forth.[9] Of course, some division of labor is possible in large-scale agricultural production, but Brewster, for example, has argued that the task and functional forms of specialization that may be possible in agriculture do not give large units a significant advantage over family-farm units, just as they were not important sources of competitive advantage for early manufacture over handicraft units: "Large units in industry finally won out only as they could and did become technologically unique through the rise of the machine process" (Brewster 1950:72; see also Braverman 1974).

Brewster contends that this technological uniqueness of large units—the use of assembly-line techniques—does not develop in agriculture because "farm operations are as widely separated by time intervals after mechanization as before" (1950:72). The introduction of machinery can reduce labor time in any one operation

9. This limitation has long been recognized. John Stuart Mill (1848:131), for example, argued that agriculture "is not susceptible of so great a division of occupations as many branches of manufactures, because its different operations cannot possibly be simultaneous."

and can transform specific operations, such as planting or harvest, into a continuous flow process. Nevertheless, the agricultural cycle remains a discrete temporal process, with sequential rather than simultaneous operations.

The second implication of the natural characteristics of most agricultural activities is that labor input is temporally uneven. With crops, certain tasks require peak labor input, while the labor requirement for other tasks is much lower; the tasks themselves may be separated by periods during which no labor is necessary. With livestock, the same pattern of variability in labor input occurs on a daily basis.

Mann and Dickinson (1978) develop a specifically Marxist analysis of the implications of this seasonality for the development of a capitalist agriculture. They argue that "the non-identity of production time and labour time characteristic of certain agricultural commodities is shown to have an adverse effect on the rate of profit, the efficient use of constant and variable capital, and the smooth functioning of the circulation and realisation process" (1978:466). This is because, in the case of many agricultural products, there is a considerable period during which "labour time is almost completely suspended as when the seed is maturing in the earth" (1978:472). These periods are "unproductive" in the sense that, although necessary to the production of the crop, they do not set labor to work and therefore do not create value for the capitalist. As a result, Mann and Dickinson conclude that seasonality poses an obstacle to the development of capitalist agriculture, and that the persistence of family farms is to be located in the resulting unattractiveness of agricultural production for capital.[10]

There is another, related implication of the seasonality of agricultural production for agrarian structure. This is the question of how a seasonally available labor force can be obtained, or, alternatively, how the production process can be organized so as to smooth over fluctuations in labor requirements. Many writers have noted the particular compatibility of the family-labor farm with the seasonality of agricultural production. It can draw on reserves of family labor during peak periods; these family members are occupied during nonpeak periods with other household reproduction

10. Echevarría (1976) makes this same point in his excellent analysis of the persistence of family-farm production (*producción parcelaria*) in Colombia in the face of the development of capitalist agriculture.

tasks, farm maintenance, or human capital formation. Core agricultural activities can be combined with ancillary activities, such as handicrafts, during slack periods. In addition, agricultural activities that are complementary in their use of labor can themselves be combined, thus smoothing the core labor requirements.

Large farms can also adopt these solutions to the problem of seasonality. The U.S. slave plantation, for example, was a carefully organized system of production designed to maximize the use of the resident (slave) labor force: the production of commodities was combined with on-farm processing and maintenance of farm inputs (Fogel and Engerman 1974). Such a closed system, in which labor requirements could be met by on-farm reserves during peak periods without generating significant labor slack during the rest of the agricultural cycle, was also characteristic of the European feudal manor and many traditional Latin American plantations. However, capitalist agricultural production approaches the problem in a different manner from traditional estate production: by the adoption of technology designed to increase the productivity of labor. This technology is generally directed first at the mechanization of the most labor-intensive and time-constrained tasks, such as soil preparation/planting or harvesting.

Many economists would theoretically locate the competitive advantage of large farms in the use of such machinery, which appears to offer these farms—but not the family farm—a solution to the difficulty of seasonal labor requirements. However, mechanization of peak tasks does not eliminate fluctuations (there are still, for example, slack periods between tasks), while it does give rise to two features of capitalist agricultural production that exacerbate the seasonality problem: specialization of production, and its spatial extension.

Specialization and physical expansion of agricultural production are logical responses to the development of specialized agricultural equipment. In a continuous-flow production process, equipment can be fully utilized by keeping it in constant operation. The sequential production process typical of most agricultural activities, however, is characterized by temporal discontinuities, so that specialized equipment lies idle much of the year.[11] This dictates the

11. Servolin (1972) points to the general problem of the idleness of (physical) capital as one of the basic deterrents to the expansion of industrial agriculture in France.

fullest possible extension of its use during its brief operating period, which means devoting as much land as the equipment can handle to the specific crop(s) in question.[12]

This specialization of the capitalist farm reduces the possibilities for smoothing labor requirements through the combining of complementary activities, and increases the absolute difference between the number of workers needed during the peak and slack periods. As Brewster argues, then, optimum farm size is determined in part by "customs which free a larger-than-family farm from labor upkeep during periods of farm unemployment" (Brewster 1950:71).[13] Social systems that provide the capitalist farms with a seasonally available work force can increase the competitiveness of the capitalist farm with the family farm, which finds it difficult to "displace" labor during slack periods. However, the family farms can, as was noted above, diversify their production in order to smooth labor usage.

The specialization and physical expansion that accompany mechanization create new difficulties for the capitalist farm. Spatial expansion of the agricultural unit may rapidly encounter managerial diseconomies, particularly if the topography or the height and density of the crop render supervision difficult. Supervision and enforcement can become increasingly costly: do you check every bushel picked, every row harvested?[14] The managerial disadvantages of large-scale agricultural production, or alternatively the ad-

12. The same phenomenon has occurred with mechanization in the U.S. dairy industry. Although machinery has been introduced that considerably shortens milking time per cow, dairy farmers have increased herd size in order to cover the costs of the new equipment. As a result, "by some inexorable working out of the laws of economics, milking time on the family farm seems to be about an hour and a half" regardless of the technology employed (Kramer 1981:38).

13. Examples of such "customs" are the parish poorhouses in early industrial England (cf. Thompson 1963), the welfare system in the United States, and various forms of "semiproletarianization" (combination of seasonal agricultural employment with family-farm production), as with temporary Mexican ("guest") farmworkers in the United States, Bantustan migrants in South Africa (cf. Wolpe 1972; Burawoy 1976), and Latin American smallholding households.

14. Johnston and Mellor (1961) discuss the difficulties of centralizing management decisions in agriculture. Rao (1971) discusses the management question in terms of "uncertainty," those activities involving few management decisions being ones with little uncertainty. Servolin (1972) argues that on the basis of the management criteria, cereal production is more easily mechanized than fruit production, and crop production in general is more easily mechanized than livestock raising.

vantages of the family farm, derive from the biological nature of that production process as well. Much more so than with industrial machinery, there is the need for constant attention to the condition and health of the crops and the livestock. The biological processes cannot be "shut off" at night and on weekends, and disease can strike at any time and spread very quickly.

Recent Marxist analyses of the capitalist labor process have emphasized the importance of the cost reductions that can be obtained in large-scale production through the application of such principles of Taylor's "scientific management" as the deskilling of labor and the payment of differential wage rates.[15] However, it is not clear that these cost-reducing practices can outweigh the managerial diseconomies encountered in many branches of agricultural production. In any case, it is important to note that many of these mechanisms of labor control and cost reduction are also an inherent part of the labor process on the patriarchal family farm. The patriarch stands at the apex of a complex age- and gender-specific gradation of production tasks among household members, with a corresponding matrix of unequal remuneration to those members. Much of the farm labor performed by younger and female household members is "deskilled," with the full knowledge of the production process vested in adult males. The ability of the patriarch to control the family-farm production process is an important element in the competitiveness of that production unit.[16] At the same time, the small patriarchal unit can avoid the problem of managerial diseconomies.

The extent of managerial diseconomies varies considerably be-

15. Braverman (1974) provides an excellent analysis of the development of this system of scientific management and its implementation in U.S. industry. I am grateful to Charles Bergquist for pointing out the applicability of these cost-reducing techniques to the analysis of agricultural productivity.

16. Wiegersma (1985) analyzes the recent resurgence of household production in Vietnam from this perspective. Her work has been very helpful to me in the development of my own understanding of the importance of the patriarchal labor process to the persistence of the family farm. In traditional African farming systems where women are the primary agriculturalists (Boserup 1970), direct patriarchal control over the production process is not an important element in the persistence of peasant production. However, while African women have retained control over subsistence production, "cash crops" throughout Africa have become the domain of male farmers, and here deskilling and differential rewards to family labor become an important element in the competitiveness of peasant commodity production.

tween different agricultural activities, rendering some more amenable to large-scale production than others. Nikolitch (1969) argues, however, that with the development of agricultural technology there is a general trend towards increasing the number of management decisions involved in any agricultural activity, contributing to the persistence of the family farm in the United States. Raup (1973:290), in a similar vein, found that a principal factor in the "flight out of agriculture" by corporations in the United States was "miscalculation of management problems in agriculture."[17]

Because of these inherent obstacles, the "industrialization" of agricultural production has been slow and uneven. It has occurred first with those agricultural products where these obstacles are "naturally" weak. Banana cultivation, for example, has a fairly stable pattern of labor requirements and a continuous production cycle, and it was therefore among the first agricultural activities in the tropics to be "industrialized." The same was true of sugar cane cultivation in some regions (Schwartz 1985:chap. 5).[18] The thrust of agricultural research in the industrialized countries has been aimed at controlling these problematic characteristics of agricultural production: at understanding and controlling the biological processes and at reducing dependence on land as a key input, with its spatial implications. The development of technology that can overcome the natural obstacles to capitalist production has, however, been more successful to date with some products than with others, and in some areas of the world than in others.[19]

## The Persistence of Colombian Peasant Production in the 1920s

This analysis provides us with an answer to the question of why Colombian peasant farmers remained competitive with large-scale

17. An example Raup gives is CBK Industries. See also Cordtz 1972.

18. Kalmanovitz (1974) differentiates among the crops grown today by the large farms in Colombia on this basis, classifying crops with fairly stable labor requirements as "plantation agriculture," as opposed to the new "commercial" subsector, where we see the greatest fluctuations in seasonal labor requirements. In the case of "plantation" agriculture, there is a stable, resident labor force, whereas in seasonal "commercial" crops there is a mobile labor force, the *iguazos*, following the crops from region to region or living in the cities during the slack agricultural seasons.

19. Mann and Dickinson (1978:472–73) point to poultry and egg production

units in the 1920s. Technology had not yet been developed or adopted in Colombia to give the larger units an edge over the inherent advantage of patriarchal peasant farms in the production of many tropical crops. This can be seen most clearly in the case of coffee production. The Food and Agriculture Organization of the United Nations concluded from the 1932 Colombian coffee census that the predominance of small coffee farms was due to the use of family labor, which gave small producers an advantage over large ones (FAO 1947). This is because traditional coffee cultivation strongly exhibits the technical characteristics discussed above. It is highly seasonal, with labor demand peaking sharply during the harvest. The harvest itself lasts for several months in Colombia, since the berries are picked individually as they ripen. Cultivation on a large scale requires large numbers of harvest workers, and generates significant supervision problems to ensure that only ripe and undamaged fruit is picked and that the leaves and branches of the plants are unharmed.

Peasant coffee producers in Colombia adopted a number of practices to deal with the fluctuations in labor requirements.[20] Coffee was rarely a family's only crop. Food crops such as maize and cassava were commonly interplanted with the coffee trees during the early years of their maturation cycle. In addition, many of the farmers in the southwestern mountains dedicated some of their land to crops that are complementary in their labor requirements to coffee, sugar cane being the most widespread. The traditional coffee trees themselves were interplanted with plantain trees for shade. Plantains were both a staple in the diet of the region and an important auxiliary cash crop. Family farms drew heavily on family and community labor during the harvests. Even toddlers were pressed into service, picking berries from the lowest branches while their older siblings and their mothers picked from the middle branches, and the adult males picked from the highest branches. Finally, familial and community relations provided the incentive to perform

as examples of the development of virtually continuous production processes, and in these activities "industrial" agriculture is most rapidly developing in the United States. On the other hand, they note the failure of experiments with the biannual production of wheat and soybeans.

20. This summary is based on the history of practices in the El Palmar region. Bergquist (1986:318–22) describes identical practices from his observations in northwestern Cundinamarca.

the harvest tasks more carefully than would be likely on estates that relied on large numbers of hired workers brought in from other regions.

As the elite attempted to cultivate coffee on a large scale, they encountered difficulties in recruiting sufficient numbers of harvest workers and high costs in the supervision of those workers.[21] As a result, the large estates began to incorporate some of the practices of the family farms, dedicating land to other crops and using plantains as shade trees (Palacios 1980:92–93).[22] More important, they attempted to incorporate the advantages of family-farm production through various forms of sharecropping arrangements. This can be seen clearly in the case of *en companía* and *contratista* arrangements described by Machado for coffee cultivation in the Santanders, and in a letter from a Cundinamarca coffee *hacendado* describing his system of sharecropping, in which the coffee trees were allocated to peasant families based on household size "so that there would be sufficient labor for the harvest" (Machado 1977:188–91, 199–201). According to this *hacendado*, his previous attempt to organize coffee production using seasonal harvest workers from Boyacá had been difficult and costly.

The system of allocating coffee trees to peasant families based on household size was also used in Tolima (Machado 1977:179–81), and a similar form of coffee sharecropping was widespread in Antioquia (Medrano 1980:72–73). In all of these cases, the peasant families also had access to plots of land for the cultivation of foodstuffs, thus smoothing over the seasonality of labor requirements for coffee cultivation. They were prohibited from growing coffee on these usufruct plots, and were sometimes required to turn over part of the usufruct output to the estate owner. They were also frequently required to contribute labor services for other hacienda tasks, sometimes receiving a minimal wage payment for that work.

Not all estates followed these practices. Particularly in Cundinamarca, many of the large estates still relied on seasonal harvest

21. Palacios (1980:95–97) provides striking figures on the seasonal variation in labor requirements on one coffee estate; this estate also encountered high costs in the supervision of the harvest workers.

22. The organization of production on the traditional coffee estates has been described in a number of recent works. See Gilhodes 1970:31–32; Machado 1977:40–52, 174–212; Palacios 1980:77–120; Medrano 1980:71–73; and Bergquist 1986:313–17.

labor in the 1920s, with all the attendant problems discussed above: costs of recruitment and transportation of harvest workers; susceptibility to rising wages, labor scarcity, or unrest; and high supervision costs. But the examples cited above show clearly that by the 1920s, the coffee *hacendados* were attempting to develop or perfect forms of production that decreased reliance on seasonal hired labor by incorporating the patriarchal family farm into the coffee estates.[23]

These various coffee sharecropping/usufruct food plot arrangements themselves began to come under pressure in the 1920s. An expanding wage labor market and rising coffee and food prices led the resident workers to begin to organize for either an increase in wages or the right to plant coffee on their usufruct plots. The *hacendados* resisted these demands (Mejía 1971:253; Gilhodes 1970:412). They saw clearly that allowing coffee sharecroppers to cultivate coffee for their own account on usufruct plots would undermine the estate's control over the labor of the household, particularly for the crucial period of the harvest.[24] Even more significantly, correspondence of the period shows that they were also worried about competition from these family farmers, who might produce coffee more cheaply and drive down coffee prices (Machado 1977:261–63). Nevertheless, *hacendados* did begin to accede to demands for more autonomy over the usufruct plots. Bergquist (1986:336) argues that the high coffee prices of the period (see table 3.1) gave the estate owners both the incentive and the ability to make these concessions. In the 1920s, then, the conditions of coffee production favored the expansion of coffee cultivation both on independent family-farm plots and on dependent family-farm plots within the confines of the large coffee estates.

23. Bergquist (1986:332) notes a trend towards an increasing percentage of permanent versus seasonal workers already evident in the coffee labor force between 1906 and 1914. According to the figures he cites, permanent workers increased by 567 percent and seasonal workers by only 240 percent over this period. It is unclear, however, to what extent this represented a shift in labor relations on the estates and to what extent it reflected the increasing weight of independent family farmers in the coffee economy, since the term "worker" is used to refer to both the estate labor force and the smallholders.

24. Machado (1977:261–63) presents correspondence from 1933 in which the coffee *hacendados* listed a number of reasons for opposing coffee cultivation on usufruct plots. Among these was the belief that it would be too costly to substitute hired labor for the harvest labor of their tenants.

TABLE 3.1. *Colombian Coffee Prices, 1916–39*

| Year | Price Index, Café Manizales[a] | Year | Price Index, Café Manizales[a] |
|---|---|---|---|
| 1916 | 62 | 1928 | 119 |
| 1917 | 56 | 1929 | 99 |
| 1918 | 71 | 1930 | 75 |
| 1919 | 120 | 1931 | 67 |
| 1920 | 93 | 1932 | 49 |
| 1921 | 67 | 1933 | 46 |
| 1922 | 76 | 1934 | 60 |
| 1923 | 82 | 1935 | 45 |
| 1924 | 111 | 1936 | 49 |
| 1925 | 121 | 1937 | 50 |
| 1926 | 124 | 1938 | 48 |
| 1927 | 109 | 1939 | 51 |

*Source:* McGreevey 1970: Table 3.
[a] New York market; 1920–29 = 100.

In the meantime, the coffee *hacendados* were becoming increasingly concerned by these difficulties. Many of them began to shift their capital into coffee processing and marketing. In 1927 they formed the National Federation of Colombian Coffee Growers (Federación Nacional de Cafeteros Colombianos), an organization that in its early years emphasized marketing rather than technological considerations.[25]

Independent peasant households also successfully competed with the food-producing haciendas throughout the mountain areas in crops as diverse as maize, beans, onions, unrefined sugar-cane products, and root crops such as potatoes and cassava. The labor process on the food-producing estates in this period has not been as well studied as that of the coffee estates, but there are indications that here, too, production frequently took place through the incor-

25. Palacios (1980: 214–22) provides a detailed analysis of the formation and early policy of the Coffee Federation, while Machado (1977: 226–38) describes the growing importance of Colombian capital in coffee processing and marketing after the 1920s.

poration of the patriarchal household system into the estate, rather than through the development of a unique form of large-scale production. This is in contrast to the specific organization of production on a large-scale basis in the cases of the Colombian banana and sugar-cane (for refined sugar) plantations in this period.

We saw that in the 1910s Aristides Benavides' parents entered into an arrangement with a *hacendado* whereby they obtained the use of a plot of land in exchange for labor on the estate. On their land they grew wheat, maize, onions, and potatoes, which they marketed in the nearby city. Although I do not have information for this hacienda on the estate products that used the labor of these tenant households, Aristides himself later worked on a similarly situated food-producing hacienda near Cali, and he provided more information on the production process at La Margarita. This estate combined cattle raising with wheat production on a large-scale basis. Wage laborers such as Aristides were employed to maintain pasture, and to clear land for wheat, sow, and harvest the crop. Aristides also mentioned crops of maize and onions. However, he did not work in these crops, and it is unclear whether they were grown on a large scale as estate crops or by tenant households.

Other studies have provided scattered descriptions of the forms of production on food-producing haciendas in other parts of the country in this period. In the upper Magdalena valley of Tolima, the decline of the tobacco industry gave rise to a resurgence of the traditional cattle latifundio, which by the early 1900s had begun to break up into medium-sized estates (Motta de Correa 1980: 131). The estates incorporated the patriarchal family-farm production system under usufruct (*arrendatario*) and sharecropping (*aparcería*) arrangements. The usufruct family received access to a food plot in exchange either for days of labor in the cattle-raising activities of the estate or payment in kind. The description of this latter arrangement is similar to the sharecropping arrangement, although here the *hacendado* generally provided tools, seeds, and other inputs in addition to the land, receiving in return a fixed share of the output.

In García Rovira province in Santander in the early twentieth century, the estates concentrated on cattle, wheat, and sugar cane (Cáceres 1980: 180–83). Wheat and cane were produced under sharecropping arrangements on small plots, while wage labor was used for some seasonal operations, such as the milling of the cane. In the late 1920s tobacco cultivation also became widespread on the estates of the region on a small-scale sharecropping basis.

As with coffee, a specific system of large-scale production had not yet been adopted for the estate cultivation of a number of basic foods. The form of production of these crops on the estates mirrored the form of production by the independent patriarchal units, but with less control over the labor process. Where the estates did opt for large-scale production, as in the case of wheat on La Margarita, small-scale producers were still able to compete successfully. In some cases, as with Aristides' parents in Cauca, small-scale production even generated sufficient revenue for the expansion of landholdings. Agricultural technology in tropical staples had not yet been shaped to the particular needs of large-scale production.

## The Agrarian Problem in Colombia

The agrarian issue of overriding concern in the 1920s was not labor militancy on the coffee estates; rather, it was the rising prices of basic foodstuffs. These rising prices were a symptom of the underlying structural weakness of Colombian agriculture, a weakness that became increasingly apparent as the booming economy generated growing demand for agricultural output. Increased production had been obtained in Colombia through the cultivation of new lands or the intensification of labor input on existing lands, rather than through the development of agricultural technology.

The possibilities for extensive expansion of agricultural production through the mountain settlement process of the nineteenth and early twentieth centuries were diminishing. While there were still large expanses of land in forest and wilderness, much of this land was held by the traditional latifundios. It appears that the price increases of the period were insufficient to convince many estate owners, in the context of rising wages and competition from small-scale producers, to expand their cultivation of basic foods. By the 1920s the ability of the agricultural sector to respond to the growing demand for foodstuffs was limited by the shortage of land for settlement, by the underutilization of estate land, by the increasing acreage being devoted to coffee cultivation, by the lack of productivity-increasing technology that could raise yields per acre, and by the outflow of rural workers to nonagricultural employment.

Several solutions to this situation were possible. One was land reform, which would open up the unfarmed "internal frontier" of the large estates to farmers eager to produce for the growing markets. This solution was adamantly resisted by the landed elite through their powerful association, the Society of Colombian Agriculturalists (SAC), which instead proposed colonization measures.[26] As noted above, however, colonization faced major economic and physical obstacles at this time. A temporary solution to the food shortage was the reduction of tariffs on imported agricultural products. This was finally implemented in 1926 with the support of some segments of SAC, particularly the large coffee growers dependent on hired labor. Imports of foodstuffs rose quickly, from 51,300 tons in 1925 to 152,100 in 1929. This measure, however, was far from a long-term solution to the agrarian problem, nor even a short-term solution, as the continued high prices of the late 1920s attest. SAC began to urge successive governments to adopt policies supporting the development of agricultural productivity. However, the events of the late 1920s intervened in the possibility of implementing any far-reaching development effort of this kind.

## The 1930s: Building Pressures as the Nation Turns Inward

The collapse of foreign trade and capital inflow in 1929 dealt a severe blow to the pattern of development, based on coffee exports and government public works expenditures, that had emerged in Colombia in the 1920s. The international price of coffee dropped sharply in 1929 and subsequent years, as can be seen in table 3.1, while foreign capital inflows all but ceased. With no financial resources to maintain spending on public works, the government at all levels drastically curtailed those projects.[27] Work on the Cali-Buenaventura highway remained suspended through the 1930s.

The industrial sector was not as severely affected by the lack of foreign exchange. By the early 1930s this sector had recovered

26. The following is based on Fajardo's (1983:36–41) overview of the shifting policies adopted in the 1920s in an effort to correct the symptoms of the agrarian problem.

27. Government spending on public works, including railroads, fell from an estimated level of 109 million pesos in 1928 to around 27 million pesos in 1930 (Bejarano 1979:237–38).

from the initial dislocation of the 1929 collapse. The cutting off of imports shifted demand to domestic producers, and the industrial sector was able to expand based on excess capacity installed in the 1920s. Industrial output in the period between 1933 and 1939 grew at an average annual rate of 10.8 percent (Mejía 1971:245–47).

However, industrial and commercial activity could counter little of the unemployment created by the cutbacks in government spending, and thousands of workers returned to the countryside in the early 1930s (Mejía 1971:259–62; Bejarano 1979:237–58). Even in El Palmar we see evidence of this reverse migration, although as a frontier community it had not been involved in the initial rural-urban migration. In the 1978–79 survey, two of the four men who had migrated to El Palmar in the 1930s came from Cali. As a consequence of the rural population influx, declining coffee prices, and the reimposition in 1931 of protectionist measures on imported agricultural products, the area sown to basic foodstuffs in Colombia increased.[28] Although the protectionist policies led to the reduction of agricultural imports, expansion of domestic production was sufficient to reduce the prices of agricultural products in the early 1930s (Bejarano 1979:245–46; Tovar 1975:37, 51).

While this decline in food prices might suggest that the agrarian problem had been resolved, both the estate and the family-farm sectors were in fact encountering increasing difficulties. Particularly in the more densely populated areas of traditional agriculture, where the public works employment of the 1920s had been an important safety valve, conflicts erupted between peasants and large land owners as the rural population swelled (Gilhodes 1970:413–14, 416–17; Bejarano 1979:278–308; Fajardo 1983:44; Mejía 1971:259–62; 1978:151; Tovar 1975:37). These growing conflicts were based in the structure of the agricultural sector, which set landless peasants face-to-face with vast expanses of uncultivated arable land controlled by the large estates.

The conflicts became most intense in the coffee zones. The collapse in coffee prices created severe financial difficulties for many of the coffee *hacendados*, who had borrowed to expand production in the 1920s. Bergquist (1986:336–38) suggests that these financial

28. Although it is likely that much of this increase in production was due to the expansion of cultivation on small and medium holdings, rather than on the large estates, this is an area in which more research needs to be done.

difficulties underlay the growing resistance by the *hacendados* to worker demands, and their attempts to reverse earlier concessions regarding the usufruct plots. The increasingly violent stance of the *hacendados* was met by the increasingly militant and better-organized stance of their workers, as both the resident labor force and the seasonal workers had been transformed by the changing economic conditions of the 1920s (the first through the increased commercialization of their usufruct plots, and the second through participation in the favorable wage-labor market of the 1920s).

The estates were caught in a double bind. Given the existing technical conditions of coffee production, reliance on wage labor would encounter significant diseconomies and would make it difficult for haciendas to compete with independent smallholding coffee cultivators. On the other hand, it was becoming increasingly difficult for the estates to maintain the sharecropping arrangements that had been developed to overcome this problem. In this context, many of the large estates began to dissolve, either directly or through government mediation parceling out their land to their resident workers (Bergquist 1986:341–42, 348; Machado 1977: 101–4). These households now joined the ranks of Colombia's independent family farmers.

In the peasant sector as a whole, however, access to land was becoming more difficult. Households attempted to maintain their livelihood through intensified use of the traditional inputs of land and labor, a process that would rapidly encounter diminishing returns on the fragile mountain soils. While access to land was becoming increasingly circumscribed for peasant producers, their marketed output was becoming ever more crucial to the nation, given the international crisis. Increased export production was needed to maintain foreign exchange earnings in the face of the falling price of coffee, and increased production of foodstuffs and industrial inputs such as cotton was needed to offset the decline in imports. Although agricultural production had been increasing in the early years of the 1930s, the intensifying rural conflict reflected the barrier the land monopoly of the large estates might pose to further expansion.

The solutions to the agrarian problem were narrowing. In 1936 the government of Alfonso López passed a major revision of Colombian agrarian property statutes, Law 200. This law was intended to resolve the dual problem posed by the monopolization and underutilization of land on the large estates. One section of the law, aimed at the "internal frontier" of the estates, declared that all land over three hundred hectares would revert to the state if left idle for ten consecutive years. The estate owners would thus be forced to cultivate their land or risk losing it to the state, which would presumably turn it over to smallholders.[29] The reforms also aimed at providing an adequate juridical and administrative structure for dealing with the rural unrest created by the premature closing of the frontier. Law 200 attempted to establish legal mechanisms for settling land disputes between estate owners and the usufruct workers and squatters on the estates.

It is important to consider this legislation's impact on the relative position of estate versus peasant agriculture. At face value, it would appear that the legislation was favorable to the peasant sector, strengthening it by increasing the access of peasant producers to land. Hirschman (1963:108), for one, has argued that this was the primary purpose of the legislation. In some areas, estate land was indeed turned over to squatters and workers. These transfers were limited, however, and there are numerous allegations that many usufruct workers were evicted by estate owners fearful of losing land.[30]

29. Since estate owners could avoid expropriation by cultivating their land, this law is illustrative of de Janvry's (1981:203–4) argument that the improvement of land use in the "non-reform" sector, rather than redistribution, has frequently been the principal intention of agrarian reform legislation in Latin America.

30. There is considerable debate over the extent to which land was transferred under the reform. Hirschman (1963:110) has argued that the majority of land conflicts were resolved through the titling of peasant operators under the law. Bergquist's (1986:341–42, 348) discussion suggests, however, that much of the parcelization of coffee estates occurred before, rather than after, the passage of the reform law, as a direct response to worker struggles. Data presented by Fajardo (1983:53–57, 60) also suggest the limited extent of parcelizations under the law (approximately 20,000 land parcels sold by the end of 1940), while pointing to the expulsion of tenants as a means of preventing parcelization. Machado (1977:320–27) discusses the expulsion of tenants after the passage of the 1936 law in some detail. The response of the landed elite, furthermore, was taking a virulent new tack. This group had always been able to call upon the local police and government officials to defend its interests against the settlers (cf. LeGrand 1980; Hirschman 1963:104). After the passage of Law 200, however, members of the landed elite from both political parties formed a violent opposition group, APEN, to defend their estates against any claims of the peasantry (Fals Borda 1969:141;

On the other hand, the estate owners did benefit from the reduction of rural tensions that followed the passage of the reform law. Despite the evictions, the creation of a legal framework within which to resolve land disputes temporarily calmed the rural unrest, as the peasants adopted a "wait-and-see" attitude.[31] I would argue that even more significant to the economic position of the estates in the long run was the impact of the legislation on their labor relations. The rural conflicts of the 1920s and 1930s had weakened the traditional labor arrangements of the estates. The 1936 legislation dealt those relations a further blow by creating the possibility that estates might lose land to workers with usufruct rights. Some writers have therefore seen the effect, if not the intent, of this legislation to have been the discouragement of traditional usufruct arrangements and the encouragement of capitalist, wage-labor relations.[32]

Nevertheless, the development of capitalist agricultural estates was still hampered by the low level of technological development in Colombian agriculture and the competitive position of the peas-

---

Mejía 1978: 162–64). This foreshadowed the methods of rural struggle that were to become endemic in the 1940s.

31. There has been much discussion of the reasons for the reduction in peasant activism after the mid 1930s. The traditional explanations have been the Hirschmanesque argument that the demands of the peasants were in fact being met by the 1936 reform; the argument that the Colombian Communist party forsook its role as leader of the peasant mobilizations in conformity with the international party position of collaboration with Liberal reforms; and the argument that the Liberal party co-opted and defused the more militant movement, UNIR, that had galvanized the rural population under the leadership of Jorge Eliécer Gaitán. But both Fajardo (1983: 53–54) and Bergquist (1986: 348–55) argue that, in fact, rural organizing efforts continued after the mid 1930s but met with little success. Bergquist provides a convincing argument, which in some respects parallels Hirschman's view, that the rural struggles of the 1920s and 1930s were motivated by a profoundly conservative desire for individual and secure landownership on the part of wage laborers as well as *arrendatarios*, *aparceros*, and *colonos*, rather than a radical vision of the transformation of conditions of work or of collective production. Once the conditions for secure title appeared to have been achieved, support for further mobilizations waned. I would add that the preservation of the patriarchal household unit as a principal form of production, within and external to the estates, reinforced this conservative focus and prevented the rural struggles from becoming more radical in their objectives.

32. Fajardo (1983: 54–56), drawing heavily on a mimeographed work by Machado, argues that this was in fact the intent of the legislation, while Bergquist (1986), in the discussion cited above, disputes this view.

ant sector in the production of coffee and tropical staples. As a result, estate owners continued to resist the mounting pressures for the transformation of their labor process and patterns of land use. Not until the 1950s, after a decade of escalating turbulence, did the transformation of estate production reach fruition. That process will be discussed in chapter 5.

In the meantime, the 1936 legislation had little effect on the conditions of the peasant producers. This sector faced intensifying population pressure on a limited land base. Chapter 4 examines the changes taking place in the family-farm economy as a result of the complex interaction of changing external conditions and the developing internal dynamic.

*Chapter Four*

# The Development of a Family-Farm Economy, 1930s–1960s

By the 1930s the children of the early settlers of El Palmar were coming of age. In order to establish themselves as family farm households, they, too, needed access to land, labor, and markets. Their possibilities were affected by broad regional, national, and even international factors shaping the development of product, land, and labor markets, as well as by factors at the local level affecting individual access to land and labor.

This chapter examines how peasant production persisted in Colombia from the 1930s to the 1960s despite the broad economic changes of the period. This persistence was a consequence of market expansion at a time when the estates were not yet able to compete effectively with peasant producers in traditional foodstuffs and coffee. The nature of that persistence changed, however, with peasant production becoming increasingly inadequate to secure the livelihood of family-farm households. Despite this deterioration, I take issue with the "vertical concentration" explanation of these economic difficulties, showing instead how the internal dynamic of the family-farm economy contributed to its decline.

Finally, I consider the issue of economic differentiation in the community of El Palmar. As we saw in chapter 1, some writers have argued that economic growth engenders a pattern of cumulative economic differences between farmers that dissolves the peasant community from within, generating an internal differentiation into a small group of capitalist farmers and a large group of landless laborers ("proletarians"). Other writers have argued that there are homeostatic mechanisms within the peasant economy that impede the development of significant cumulative differences between households. I demonstrate that both cyclical and cumulative factors generated economic differences between El Palmar households, with cumulative differences becoming more important as local conditions deteriorated.

## The 1930s: Market Production and Access to Land

Colombian coffee output continued to increase during the 1930s, despite the depressed world market prices evident in table 3.1. Increased production in the face of this decline has been attributed to three factors: the maturation of coffee trees planted in the boom period of the 1920s, which were reaching their peak production levels in the 1930s; currency devaluations, urged by the Coffee Federation, that in part countered the falling international price of coffee; and the nature of family farming, which enabled smallholding coffee producers to withstand price declines (Palacios 1980:214–21; Bergquist 1986:308).

Although the maturation of coffee trees did contribute to increased output, production increases were also due to the physical expansion of coffee cultivation to new lands (Bergquist 1986:308). It is likely that, given the difficult economic and social conditions on the coffee estates, this physical expansion was taking place primarily on family farms.[1] In El Palmar, farmers acquiring parcels of land in the 1930s began to plant new coffee on portions of 30 percent of those parcels.[2]

Despite currency devaluations, the internal price of coffee was depressed in the 1930s. How, then, do we account for the continued physical expansion of coffee cultivation by smallholders?

1. Bergquist (1986:300) concludes from a comparison of the 1932 and 1939 coffee censuses that the "number of small coffee farms mushroomed" during this period. The failure in these early censuses to indicate tenure status of the coffee operations, however, means that we cannot distinguish owner-operators from the sharecroppers on the estates.

2. These and subsequent figures on 1930s production in El Palmar come from the 1978–79 household survey.

The "nature of the family farm" interpretation of this phenomenon is that falling coffee prices squeezed coffee smallholders, now dependent on the market for many of their necessities, and forced them to expand cultivation of their cash crop in an effort to maintain their standard of living. They carried out this expansion by intensifying their labor effort, ignoring the declining labor productivity associated with this intensification (something that employers of wage labor could not do).[3]

While there is some truth to this argument as a description of how coffee cultivation was expanded in this period, I believe that it paints too negative a picture of the conditions of the coffee smallholders. I would argue that, both because of the relatively recent settlement of coffee lands and because coffee cultivation fit well into a mixed cropping pattern suitable to the organization of the labor process on the family farm, these farmers expanded the cultivation of coffee, not out of desperation, but rather because it continued to give them a good return.[4] At the same time that they ex-

3. Palacios, for example, argues that "marginal productivity was hardly of major concern for the small-holding peasant" (1980:228). Bergquist (1986:308) provides a slightly different perspective, presenting the argument of the president of the Coffee Federation in 1934, Mariano Ospino Perez, that peasant producers could withstand declines in the price of coffee because of their production of subsistence crops. This would not, however, explain their *expansion* of coffee production while prices were declining.

4. Many writers have depicted the conditions of the small coffee farmers in extremely negative terms, not only for the period of the thirties but more generally, ascribing their difficulties to the process of vertical concentration of coffee production through which they were squeezed by coffee processors and merchants. This argument has been made, for example, by Machado (1977) and Fajardo (1983). Bergquist (1986:311–12), who depicts the conditions of the smallholding coffee cultivators in more favorable terms, also argues that the victories of the coffee workers in gaining access to land were illusory, since they were "only to be exploited more efficiently and easily through capitalist control over coffee commerce" as domestic capitalists developed a "monopoly over the means of coffee exchange." Palacios (1980:243), on the other hand, points out that the income of small coffee farmers has been protected by the favorable internal support prices set by the Coffee Federation. This argument is consistent with the recent experience of the farmers in El Palmar, as I shall show subsequently—although it is also true, as Palacios demonstrates, that the lion's share of the coffee income has gone to the Coffee Federation. There is some evidence that the general prosperity of family farmers in newly settled coffee zones such as Dagua was not that different from the status of family farmers in other recently colonized areas of the country. Writing in 1935, a traveler through the north coast department of Bolívar reported that family farmers in the *sabana* zone, which had been colonized beginning in

panded coffee cultivation, they also expanded the complementary complex of staple foods and sugar-cane production as well—again, despite falling prices of foodstuffs. In El Palmar, farmers planted permanent crops other than coffee (primarily sugar cane and plantains) on portions of 40 percent of the land parcels that they acquired in the 1930s. In addition, they reported planting annual crops such as maize, beans, and cassava, both for market and for household consumption, on portions of 50 percent of the new parcels.[5] Much of the land they acquired in this period, particularly through inheritance, was still virgin. Land transactions therefore led to increased total acreage under cultivation in the region, as some of the newly acquired land was converted from forest, pasture, and *rastrojo* (undergrowth on idle land) to crop land.

This is not to suggest that there were no storm clouds on the family-farm horizon. There were indeed. These clouds were not formed by falling product prices, however, but rather by mounting population pressure on a limited land base; and while these clouds posed only a distant threat to family farmers in newly settled areas, they were already enveloping other peasant communities. We saw in chapter 2 that population pressure was already a problem in the 1800s in some highland regions. Farmers in these areas were reaching the ecological limits of existing resources, given the fragility of the mountain soils and the dependence of slash-and-burn agriculture on fallow periods for regeneration of soil fertility. These regions were therefore the source of permanent migration to colonization zones, as well as temporary migration in the late 1800s and early 1900s to meet the seasonal labor demand of the great Cundinamarca coffee estates. Finally, in the 1920s, migrants from these areas filled the ranks of the wage-labor force in public works and industry.

In the 1930s all these income-earning possibilities for the surplus population of the highland communities were evaporating, while declining product prices reduced the income-generating po-

---

the nineteenth century, were still fairly well off. They had diversified agricultural operations, with sugar cane as an important market crop, and most owned *trapiches* (Diaz 1935, cited in González 1980: 101–2).

5. The planting of annual crops was probably more widespread than these figures indicate, since the farmers tended not to mention annual crops planted primarily for home consumption.

tential of the existing land base. These pressures forced migrants into the few remaining areas of the country where climatic conditions and population/land ratios were still favorable, as in the Dagua mountains. At the same time, the dynamic of large family size was also contributing to rapid population growth in these frontier regions. Appendix A shows that the population of the *municipio* of Dagua increased from 2,331 in 1905 to 3,146 in 1912. By 1938 it had mushroomed to 14,149, to become one of the larger *municipios* in Valle del Cauca. Large family size, while a fundamental pillar of the patriarchal farm economy in this period, was beginning to generate potentially serious limits on the ability of this family-farm system to reestablish itself in succeeding generations.

This difficulty had not yet reached serious proportions by the 1930s in El Palmar. Despite the falling prices of foodstuffs in the cities of the nation, the combined cropping system of the established farmers like Juan Álvarez and Aristides Benavides continued to generate sufficient revenue for them to expand their holdings in El Palmar and in neighboring areas. Land-poor households—both those headed by the sons of resident farmers and those of new arrivals in the community—also encountered a variety of means by which they could accumulate funds to acquire land. Various *en compania* arrangements through which they farmed the "surplus" land of established landowners (often their own fathers or uncles), agricultural wage labor, and agricultural trade were all important means of building a land fund. The continued growth of Cali also generated income-earning opportunities for some aspiring landowners.

The stories of Lubino Cabrera and Agapito Cabrera, two later arrivals to the area whose background was discussed in chapter 2, illustrate the way in which regional growth in this period continued to provide an advantageous environment for migrants to the mountain zone.[6] Their stories also, however, illustrate the difficulties that some men were beginning to encounter in their quest for land ownership. Later arrivals found a tightening land market, as they competed with the offspring of the early settlers for access to the

6. Information in this chapter on Agapito and Lubino Cabrera is based on oral histories recorded with Ursulina and Lubino Cabrera and members of their family in 1978–79 and 1983–84, and on the interview data for these households.

land controlled by those patriarchs. Increasingly, a migrant's success in El Palmar became related to the financial resources at his disposal when he arrived in the community.

Agapito decided to move to Cali from Nariño in the late 1920s. He managed to get a contract to feed the workers in the large Golondrinas coal mines outside of the city. Lubino and Ursulina decided to follow Agapito to Cali. Lubino worked in the mines, and Ursulina worked with Agapito's wife cooking meals for forty to fifty mine workers. With this contract, Agapito was able to save a considerable amount of money. He began to work *en companía* with a landowner in El Salado, growing foodstuffs for the Cali market. Lubino, too, began to work in El Salado.

In the early 1940s agricultural production in the El Salado region was given a boost by resumed construction of the Cali-Buenaventura highway, itself stimulated by the continued growth of the coffee trade in the 1930s. By this time the Anchicayá advocates had gained the ascendancy, and the route was redirected west from El Treinta, passing through El Salado and El Queremal on the way to Buenaventura (see map 4). The farmers in the region found another important market for their products. Aristides recounted how he, along with other farmers who owned land around El Queremal, sold *panela* on contract to a middleman purchasing provisions for the workers. Aristides received seventy pesos per *arroba* (25 lb. load) of *panela*, compared to the twenty pesos he was receiving before construction started.[7] As he describes it, "then the price of *panela* rose, and the price of everything rose because people came like ants, to work [on the highway]."

Agapito prospered with his *en companía* production. With the income from these crops, he was able by the early 1940s to buy a large farm in El Palmar: sixty-three acres, for which, according to Ursulina, he paid six thousand pesos. By the mid 1940s Agapito had much of his land in coffee and employed thirty to forty workers during the harvest. From the vantage point of the land prices of 1979, Agapito's purchase price seemed to Ursulina like a gift. However, it was high enough to put land out of Lubino's reach. Lubino had not been as successful as his brother in building his savings in the mines of Golondrinas and on the farms of El Salado. In the mid

7. One Colombian pound equals 500 grams.

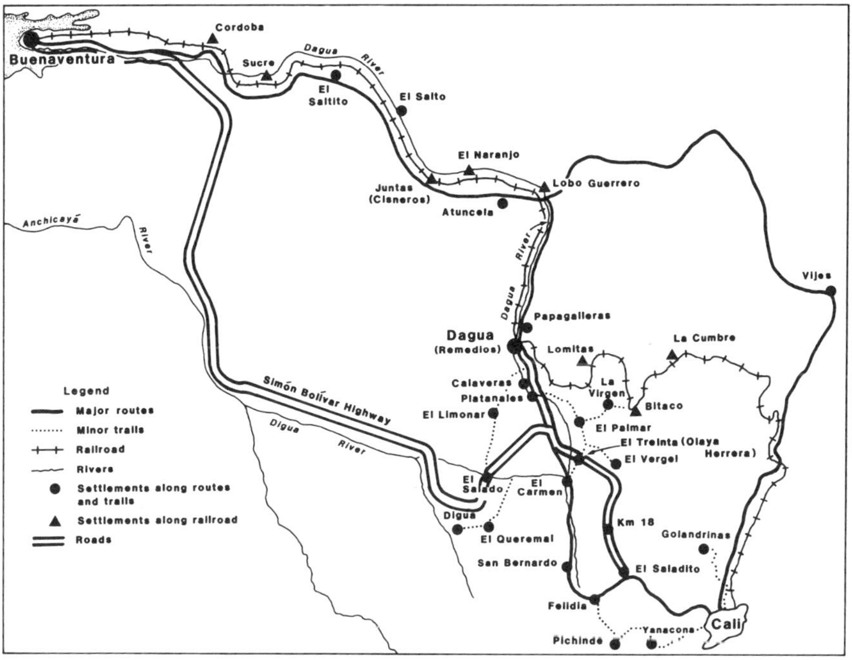


Map 4. The Simón Bolívar Highway through the Dagua Mountains (the Anchicayá Route), 1946.

1940s Lubino followed Agapito to El Palmar, where he, his wife, and two of his children began to work *en companía* with Agapito picking coffee.

## The 1940s and 1950s: National Economic Growth and Land Scarcity

After World War II the international economic recovery spurred rising world market prices for coffee (table 4.1), which gave rise to another growth phase in Colombian coffee production.[8] The Dagua

8. Coffee prices had actually begun to rise even earlier in Colombia. As a means of obtaining "Pan American Solidarity" against the Axis nations, the United States, by then the major purchaser of Colombian coffee, had entered into the InterAmerican Coffee Agreement in 1940 to aid Latin American producers who had a severe excess supply of coffee. In Colombia, the Fondo Nacional de Café was created to finance the purchase and warehousing of coffee through special

TABLE 4.1. *Colombian Coffee Prices, 1940–61*

| *Year* | *Price Index, Café Manizales*[a] | *Year* | *Price Index, Café Manizales*[a] |
|---|---|---|---|
| 1940 | 37 | 1951 | 254 |
| 1941 | 64 | 1952 | 247 |
| 1942 | 69 | 1953 | 260 |
| 1943 | 69 | 1954 | 347 |
| 1944 | 69 | 1955 | 280 |
| 1945 | 70 | 1956 | 322 |
| 1946 | 98 | 1957 | 278 |
| 1947 | 132 | 1958 | 227 |
| 1948 | 142 | 1959 | 197 |
| 1949 | 163 | 1960 | 195 |
| 1950 | 231 | 1961 | 190 |

*Source:* McGreevey 1970: Table 3.
[a] New York market; 1920–29 = 100.

farmers continued to expand their cultivation of coffee,[9] as did mountain farmers throughout the country. The coffee censuses of 1932 and 1955 show that the number of coffee farms in Colombia increased from 149,206 to 234,674 over this period (Palacios 1980: 229, table 31). It is likely that much of this increase was due to the breakup of family farms through inheritance, as was the case in El Palmar, with each inheritor planting his or her own plot of coffee. Although the mean size of these coffee plantings increased over this period, by 1955 it was still only 3.3 hectares for the country as a whole, and as low as 1.4 hectares in some departments. At mid-century the techniques of coffee cultivation were virtually identical to those in use at the beginning of the century, and smallholding

taxes on coffee exports. By 1941 internal coffee prices were rising (Palacios 1980: 220–21).

9. The Coffee Federation worked actively with the family farmers of this region in the postwar period. In El Palmar it helped farmers to build coffee-drying units and to adjudicate land titles, as well as building an aqueduct in La Virgen that served both communities (based on interviews with El Palmar residents in 1978–79).

coffee cultivators continued to compete effectively with large-scale producers (Palacios 1980:228–29).

Industrial growth in the postwar period was exceptionally high. Between 1945 and 1950 the average annual industrial growth rate was 11.5 percent, while from 1949 to 1956 it averaged 8.2 percent (Mejía 1971:249; Fajardo 1983:71).[10] The nation's population grew ever more rapidly: 2.2 percent per year between 1945 and 1953, and 2.7 percent between 1951 and 1964. But the growth of urban areas, as a result of accelerating migration from the countryside, reached much higher levels: an average annual rate of 5.2 percent between 1945 and 1953, and 5.6 percent between 1951 and 1964 (ECLA 1956:17, table 4; Inter-American Committee on the Alliance for Progress 1968:82).

The economic center of the country continued to shift westward, as coffee and industry burgeoned in the mountains and cities of the Antioquia and Valle del Cauca regions. Coffee exports through Buenaventura increased steadily through the 1940s and early 1950s. Antioqueño coffee, which had been going out by the Magdalena River to the Caribbean coast, began to be shipped out through the Pacific after the completion of the Antioquia-Cali railroad line in 1942 (Parsons 1968:145–46). Cali continued to thrive as a center of western trade and commerce, as well as an important industrial site.[11] Its population also continued to grow rapidly, as Appendix A demonstrates, at a phenomenal average annual rate of 13.5 percent during the 1940s. While the population of Cali almost tripled be-

10. During World War II industrial growth had slowed due to a shortage of needed imports for a sector now operating at capacity. The average annual rate of growth of industrial output was only 3 percent between 1939 and 1945 (Mejía 1971:249). The shortage of imports, however, had enabled the country to build up large reserves of foreign exchange from coffee exports. Once the war ended, these reserves and the strong market for coffee funded the imports needed to support a second wave of industrial expansion in Colombia.

11. According to Banderas, writing in 1944, Valle del Cauca was one of the most important commercial centers in the country because of the shipment of merchandise from Buenaventura through the Panama Canal. Cali was the principal distribution center for the imports and exports of several western departments. In addition, it was an industrial center of some significance. Banderas details the industrial development of the region, including the processing of agricultural and other primary products, agricultural inputs (pharmaceuticals, coffee depulpers, mills), light consumer-goods production, and a growing machine-repair industry (1944:382–84). A DANE study (1955:19) concludes that by 1940 Cali had become the "principal economic center of the southwest."

tween 1938 and 1951, the population of Valle del Cauca almost doubled in that period, to 1,106,927 in 1951. In the 1950s Cali continued to grow at the amazing rate of 13 percent per year, reaching 464,800 inhabitants by 1958.

### *Local Conditions: Commercial Prosperity and Growing Land Scarcity*

This postwar boom generated extremely favorable market conditions for the farmers in the Dagua mountains. The completion in 1946 of the Simon Bolívar highway (Silva Holguin 1960:234), as the new Anchicayá route between Cali and Buenaventura was called, allowed the trip between those two cities to be made in seven hours. Along with the Cali-Dagua road, the new highway greatly increased traffic through the mountain area and stimulated local and regional trade. In addition to traditional foodstuffs, many of the farmers of El Palmar began to grow pineapples for the expanding urban market. The new highway also shifted marketing patterns within the mountain region. Dagua lost its importance as the principal regional market, although it retained its role as a major railroad terminal and a local market. For El Carmen, as well, this new highway initiated a decline. Market activity soon moved to the growing settlement of El Treinta, the crossroads of the new highway system (see Map 4). This was a marked improvement for the farmers of the El Palmar area, as El Treinta was only a few kilometers away. The residents of El Palmar joined together in the late 1940s to build a road to this new market town. El Queremal also became an important market town during this period, attracting farmers from the Digua River area.

The expansion of agricultural markets and rising agricultural prices benefited land-poor households like the one headed by Lubino Cabrera. With the income from his family's *en companía* arrangement with Agapito, Lubino was able to accumulate enough money to buy 7.5 acres of uncultivated land in 1948. This was not sufficient to support Lubino and Ursulina's large family—they had eight children by the 1950s—but they were able to augment farm income from a variety of other sources in the favorable market conditions of the 1950s. They continued to work with Agapito, and also *en companía* with some of the large landowners of the area. They grew cassava for the El Treinta market, selling to buyers who bought the cassava by lot and harvested it themselves. Lubino also bought pineapples from his neighbors to resell in the *galería* in Cali. Through these diverse livelihood strategies, the Cabrera household was able to support itself through the 1950s, but Lubino was unable to accumulate sufficient funds to purchase more land.

The children of the earlier settlers in the area were encountering similar circumstances. While recovering coffee markets and the urban-industrial expansion of Cali provided them with expanding market opportunities, farm land was becoming more limited. While they, unlike Lubino, could expect an inheritance in the area at some future date, they were faced in the meantime with fewer opportunities for land purchase than their fathers had encountered.

## *Processes of Adaptation to Diminishing Farm Size*

Types of crops and cultivation practices began to change in El Palmar as farmers adjusted to market conditions and smaller farm size.[12] For those who inherited or purchased smaller plots, the amount of land they could put into sugar cane could not justify the cost of a *trapiche* to process it. Other basic food crops also became less feasible as farm size declined. While crops such as cassava, *arracacha*, or maize required very little labor or other inputs, yield and market price were such that the cash return per acre was low. Land, not labor, was now the scarce factor of production.

The small farmers therefore increasingly concentrated on market crops with high value per acre, such as coffee and pineapples, with small plots of foodstuffs primarily to meet household needs. An important component of coffee cultivation was the use of plantains as shade trees for the coffee plants. They produced year round and provided the farmers with an essential food as well as a supplemental source of cash. One modification to traditional practices that became widespread during the 1950s helped the farmers of El Palmar to counter the trend towards declining farm size. This was the mixed cultivation of maize and kidney beans. On small farms the maize was primarily for household consumption, while the

12. The discussion of farm practices in this section is based on the interviews carried out with 105 resident households in El Palmar in 1978–79.

beans provided a source of cash income. Another response to declining farm size was to farm the land more intensively, leaving less land fallow, cultivating more marginal land, and planting more permanent crops.

With these modifications in land use and farm practices, in the context of the favorable market conditions of the 1950s, the farmers of El Palmar were able to stave off the impending crisis of the deteriorating land base. Thus, as Chayanov postulated (1966: 113–15), land scarcity in the family-farm economy can be countered by a shift to labor-intensive practices that increase the value of output per acre, provided market conditions make such a shift feasible. The intensification of land use through the 1940s and 1950s was not without cost, however. While they provided greater income per acre in the short run, these agricultural practices contributed to soil deterioration through erosion and depletion of nutrients. The planting of maize with nitrogen-fixing beans, and of coffee with plantains and *guamo* (*Inga* spp.), a nitrogen-fixing tree, helped to alleviate, but could not counter, this trend. Pineapples, *arracacha*, and cassava are crops that draw large quantities of nutrients from the soil, and increasingly these nutrients were not being replenished.[13] At the same time, the farmers were becoming increasingly dependent on market conditions for a narrow range of crops suited to their land-scarce environment.

## *Processes of Differentiation in a Family-Farm Economy*

Over the period under consideration, two trends were emerging in this family-farm community. One was the tendency, as a result of the patriarchal logic of large family size, for holdings to be fragmented through inheritance. Average farm size therefore declined. The other was the tendency for the economic status of the households to vary sharply around that declining trend—as evidenced, for example, by the difference between Agapito's 63 acres and Lubino's 7.5 acres in the 1950s. The presence of either of these tendencies—shrinking average farm size or economic differences between

13. This was the conclusion reached by farmers in El Palmar, reported to me during interviews in 1978–79.

peasant households—has been taken in studies of the "peasant question" as evidence of the disintegration of the family-farm sector under the pressures of capitalist development. Was this taking place in El Palmar? Were expanding markets enabling a few farmers, such as Agapito, to prosper, while gradually forcing the majority to rely increasingly on wage labor?

The pattern of changing farm size in El Palmar was more complex than this simple differentiation model suggests, responding to an interplay of both external and internal factors. At least four distinct processes accounted for the observed differences in farm size between households at any given point in time. Heads of households differed in the extent of their accumulated holdings because they were at different stages in the life cycle, because of differences in their parents' economic status, because of differences in the way the offspring of any given family had been treated by the patriarch, and because of changing market conditions, which helped or hindered households at some stages of the life cycle more than others. In addition to these systematic differences, there were variations due to random factors such as personality or luck.

*The Family-Farm Life Cycle.* Chayanov (1966) developed a theory of the "family-labor farm" as a unit of production whose scale responded in an elastic manner to the reproduction (consumption) needs of the farm family and the availability of family labor. In this analysis, farm size would be increased as household consumption needs and the family labor force grew. By the time the male head of household was in his forties, the offspring would be reaching working age and the household would be expected to be consolidating and expanding holdings acquired earlier. The farm and the family labor force would be at their maximum size by the time the head of household reached his fifties. They would both begin to decline thereafter, as the offspring matured and formed their own households and farm size was gradually reduced. Although there would be minor variation in farm size among households at each stage because of differences in subjective valuations of consumption utility and work drudgery, this basic pattern would repeat itself over and over again, with no trend towards concentration of holdings. The Chayanovian world is one of rhythmic cycles of growth and decline within an overall context of balance and stasis.

Clearly, the farmers of El Palmar have not lived in a pure Chayanovian world. Chayanov did not consider the possibility of using hired labor, which would free the family from the drudgery limitation on expanded production. Nor does his model of self-exploitation reflect the internal hierarchical relations of the family farm, in which the investment and consumption decisions of the patriarch can be related to an increase in the drudgery of subordinate household members rather than an increase in his own labor input. Growth of the family farm may thus proceed far beyond the size commensurate with the basic consumption needs of the family, as was the case with many of the early farms in El Palmar.

Nevertheless, we should not reject Chayanov's model out of hand. A pattern is discernible in El Palmar with regard to farm size and family life cycle that resembles that postulated by Chayanov. The data in table 4.2, which include only male-headed households, reveal this pattern.[14] The age of the head of household is a good proxy for stage in the family life cycle, as there were statistically significant differences in household composition related to this variable. These stages were characterized by a progressive expansion and subsequent contraction of the number of household members, paralleled by changes in the number of dependent children and of working-age members.

The relationship between the age of the male head of household and the amount of land owned conformed, with one important exception, to the expected Chayanovian pattern of gradual increase and then, for the oldest group, decrease. The exception to this pattern was the cohort of men in their forties, whose average holdings were smaller, not larger, than the holdings of men in their late thirties. Not only does the experience of the forties age group belie the universal appropriateness of Chayanov's model, but the experience of the households overall conformed only weakly to the Chayanovian pattern. While there were fairly sharp differences in mean farm size among the five cohorts, these differences are only statistically significant at the 80 percent level. This indicates that, within each age group, there was considerable variation in farm size that

14. Female-headed households in El Palmar would be expected to have a different dynamic, since they consisted principally of young mothers with little or no land, whose primary means of support was remittances from the father of their children; or of older widows whose land holdings ranged from minimal to sizeable.

TABLE 4.2. *Family/Farm Life Cycle in El Palmar, 1978–79*

| *Age of Head of Household* | *Mean Household Size*[a] | *Mean Number of Children under 10*[b] | *Mean Land Owned (acres)*[c] |
|---|---|---|---|
| 25–29 | 4.4 | 2.1 | 5.0 |
| 30–34 | 5.3 | 2.0 | |
| 35–39 | 6.2 | 3.2 | 7.7 |
| 40–49 | 7.6 | 1.9 | 5.8 |
| 50–64 | 6.2 | 1.7 | 28.2 |
| 65+ | 4.8 | 0.9 | 18.3 |

*Source:* Author's interviews with 88 male-headed resident households.
[a] Analysis of variance significance at 99 percent level.
[b] Analysis of variance significance at 92 percent level.
[c] Analysis of variance significance at 80 percent level.

overlapped the differences between age groups. To what were these differences due?

*Family Background and Patriarchal Relations.* In the context of diminishing access to land, we might expect those men whose fathers had substantial holdings to be in a better economic position over the course of their life cycles than the sons of land-poor patriarchs. Although this was indeed a contributing factor to economic differences in El Palmar, it was not determinate in the experience of the cohort of men who were becoming heads of households in El Palmar in the late 1940s and early 1950s. Seventeen survivors of this group of men—those who had not died or left the village—were interviewed in 1978–79, when they were between the ages of fifty and sixty.[15] By 1978–79 the accumulated holdings of these men ranged from 1.5 acres to 90 acres. Thirty percent had acquired "large" farms (over 22.5 acres), 41 percent had acquired

15. The average age at marriage for these men, estimated from the age of the oldest living child, was twenty-six (this method, however, contains an upward bias). They were offspring of earlier settlers, or had migrated to the region as young men in the 1930s or early 1940s. I am not including here eight men in this age group who migrated to El Palmar in the 1950s or later, as this is a diverse group of men who arrived at different times and ages and with different resources at their disposal.

"medium" farms (7.5 to 22.4 acres), and 29 percent had only acquired "small" farms (under 7.5 acres).

These differences were by and large unrelated to family background (as measured by the size of the parents' holding, one of the determinants of the size of the descendant's inheritance). A number of descendants of land-poor families had risen into the ranks of "medium" or "large" peasant farmers in the 1950s through land purchases. These purchases had been financed through such economic activities as sharecropping, agricultural wage labor, and agricultural trade. Others did not experience much economic mobility in the 1950s and 1960s, but were in a position to improve their economic status in the 1970s by taking advantage of the new opportunities of that decade. On the other hand, descendants of landed patriarchs in El Palmar were not necessarily the men who were most able to build their own farms over the years.

Why was family background not an important factor in the ability of these men to establish themselves as landed patriarchs in El Palmar? Two other factors had vitiated the impact of differences in family background: the nature of relations between landed patriarchs and their offspring with respect to access to land and other resources, and the existence of local off-farm opportunities for the economic advancement of the offspring of land-poor households.

Partible inheritance has been the rule in Colombia since colonial times. However, the patriarch has full authority over the disposition of his resources during his own lifetime. He can sell off his property if he so desires: both Juan Álvarez and Aristides Benavides sold landholdings in various regions, although they had grown children by that time who might have wanted to farm that land.[16] Juan, in particular, sold off much of the land he had amassed in La Virgen and elsewhere.[17] If his land had been subdivided among his ten

16. Material in this section on the Álvarez and Benavides families is based on oral histories recorded with Juan Álvarez, Aristides Benavides, and members of their families in 1978–79 and 1983–84, and on the 1978–79 interview data for these households.

17. Some of that land was donated to build the La Virgen school. The bulk of it was sold in parcels to neighbors. With the proceeds Juan bought property in El Treinta. In the 1970s Juan sold the land in Digua and bought a house in Cali. When Juan's wife died, and Juan became ill in the late 1970s, the property in El Treinta was sold and he moved to the house in Cali with his youngest daughter. Yolanda cared for him until his death in 1982, when she inherited the Cali property.

children in the early 1950s, each would have received fifteen acres in La Virgen as well as land in Digua. When the farm was finally divided in 1978, each son and daughter got approximately six acres in La Virgen and no land elsewhere.

Proceeds from land sale, as well as proceeds from farm and other operations, can be spent by the patriarch on his own and his household's consumption. However, these proceeds can also be used by the patriarch to improve the independent economic status of his offspring, providing them with education or setting them up in their own farms or businesses. The independent economic success of the patriarch's offspring is a means of attempting to assure his own support in old age, and would also be seen by his wife as a useful investment in old-age security.[18] Because these types of investments represent a transfer of resources to the offspring, however, they pose two difficulties for the patriarch. First, there is no guarantee that the offspring will recompense their parents later in life. The patriarch would therefore want to be selective in his choice of beneficiaries, giving the most assistance to those offspring who had been most obedient or with whom the parents had the strongest ties. Second, the independent economic activities of the offspring represent a loss of labor power for the patriarch's own farm operation. This is particularly a problem for those patriarchs with sizeable holdings.

Rather than setting up his offspring independently, then, the landed patriarch may attempt to keep them on the farm. His ability to do this is constrained by the existence of off farm opportunities. The greater these opportunities, the more the patriarch may need to offer to keep his offspring available. He may have to pay them for their labor, give them greater decision-making authority, give them greater direct control over resources through some form of sharecropping or rental arrangement, and perhaps even give them part of his holdings as an "early" inheritance. Given the subordinate role of females in the patriarchal households and in the agricultural operations of the Colombian mestizo communities, it is likely that daughters would receive the fewest concessions of this kind from the patriarch.

These considerations suggest that, even in the case of partible

18. Caldwell (1982) in particular has presented a persuasive analysis of patriarch-son relations as old-age security strategy.

inheritance, the extent to which offspring benefit from their parents' property is highly variable. It depends on the manner in which individual patriarchs choose to respond to these various pressures. We can see how these factors operated in the treatment of their offspring by two landed patriarchs, Juan Álvarez and Aristides Benavides.

We have seen that Juan sold off much of his land. He used the proceeds to buy several houses in La Treinta, as income property and a residence for himself and his wife. However, he retained sizeable holdings in the La Virgen area and relied heavily on the labor of his children to farm that land. His relations were most difficult with his oldest son, Miguel. As Miguel matured, Juan tried various methods of keeping him on the farm, but Miguel eschewed the *en companía* arrangement offered by his father, sold off the small plot Juan finally gave him, and left with his wife to wander around the region as a day laborer. The other six sons were more responsive to Juan's efforts, entering into *en companía* arrangements with Juan as they matured. Two of the sons remained at home under these arrangements, operating as a semi-autonomous labor force. They did not purchase land of their own, and did not marry until quite late. With the other four sons, however, Juan's *en companía* arrangements had the paradoxical effect of enabling them to gain full economic independence from the family unit. They were able to accumulate funds with which to set themselves up independently—two as local farmers, one as a farmer in another community, and a fourth as a trucker operating out of El Treinta.

If we looked at the economic status of these men at a particular moment in time in the late 1950s, we would see differences in economic status that were clearly unrelated to family background, and only in some cases related to differences in the stage in the life cycle. The oldest son was the poorest. Already in his early forties, and with seven children, Miguel had just returned to the area after many years of wandering. With no land of his own, he lived in one of his father's houses in La Treinta and worked as an agricultural laborer. His older children also worked as *jornaleros*. The second son was still in La Virgen, working his own small plot of land and also working *en companía* on his father's land. Juan's third son lived in El Treinta, earning his living transporting produce in his truck. These two sons were in their late thirties, married, and with growing young families. Their father's resources had given them the means to earn a livelihood, but neither had yet become wealthy in his own right. The next son was unmarried and living at home, working *en companía* on his father's land. The three youngest sons, recently married, were also working *en companía* with Juan. Only one of these four men, all in their thirties, had so far been given land of his own.

The same pattern of differential treatment can be seen in the case of Aristides' household. His oldest son, Manuel, received little education. When he matured, Aristides did not give him land of his own, because, he explained, Manuel had a tendency to wander and had never married. Manuel worked on Aristides' farms, sometimes in El Palmar, sometimes in El Queremal, but he would periodically take off and travel around the country as a day laborer. It is noteworthy that the same "wanderlust" was evidenced by Juan's oldest son. It was suggested in chapter 2 that one of the motivations for the nineteenth-century migration of young males in this region was the "push" stemming from the patriarchal relations of the households of their origin. These pressures may be strongest in the case of the oldest son, and, not surprisingly, in both households the oldest son became a "wanderer" (as well as an early drinker).

Aristides' next two sons fared only slightly better. They received a basic education (three years of elementary school) by sporadically attending classes, but farm work always came first for these oldest male children.[19] As they matured, they worked *en companía* with their father. When they married in the early 1950s, Aristides gave them each a small plot "in order for them to continue and to build their little houses." One, Ricaute, received three acres of land, half of which was in coffee. The other, Arcesio, was given two acres, one in coffee, and he and his bride Diomelina (one of the daughters of Juan Álvarez) were given a part of the main house to live in.[20]

19. Aristides finally started a school in his own house in El Palmar in the late 1930s, with a teacher who came from Dagua, to make it easier for the children to combine school attendance with farm work.

20. The entrance of Arcesio's wife, Diomelina Álvarez, into the household made it easier for Aristides to maintain farms in both El Queremal and El Palmar while also educating his daughters. The daughter-in-law took on most of the responsibilities for the El Palmar household. Diomelina recounted that when she arrived, two of the brothers were still living in the main house (Manuel and one of

Thus Aristides also followed a strategy of combining *en compañía* arrangements with transfers of land to his sons—but he gave them amounts of land that were insufficient for their needs. These two sons therefore remained available for some time as workers on their father's farm. Thus the patriarch is seen to tread a fine line between granting the sons enough authority and income to keep them satisfied and not giving them so much that they can become completely independent.

The two youngest sons fared differently. The youngest was sent to study "various artisan crafts" in Cali—Aristides' attempt to take advantage of the changing economic opportunities occurring as a result of the postwar boom (Aristides also had his youngest daughter trained to be a seamstress). When this son married, Aristides sold him a house that he owned in Cali. The second youngest son, Alfredo, lived and worked with his father until he married. Aristides then gave him a 50-acre farm in El Limonar. Why, Aristides was asked, had he given Alfredo a 50-acre farm when Ricaute and Arcesio had received only a few acres each? Aristides explained that a friend had offered him the farm for a good price (thirty thousand pesos) and he wanted to buy it. Alfredo had just married and needed some land. Aristides wanted to help him because he felt that Alfredo was a good worker (an example of the patriarch making his investments where he hopes they will give the greatest return). Aristides decided to buy the land in Alfredo's name, giving him the title, with the understanding that Alfredo would no longer have any claim to Aristides' property ("ya usted está afuera de parte"). This would be Alfredo's inheritance. Aristides felt that this was fair because he had five hundred acres in El Queremal that would be divided among the other children. Be that as it may, the El Queremal inheritance was far in the future (in 1983 Aristides still had not subdivided that property). In the meantime, Alfredo farmed his fifty acres in El Limonar, while Ricaute and Arcesio struggled with their small plots in El Palmar.

Again, if we were to look at the status of the sons of Aristides

---

the younger sons). The parents spent most of their time in El Queremal, and the other children were either with them, away at school, or married. Diomelina cooked for the brothers-in-law as well as for her own family, and later for some of her older sister's children who were studying in the El Palmar school.

Benavides in the late 1950s, we would see differences unrelated to family background, and even less related to stage in the son's life cycle than in the case of the Álvarez offspring. Only one son had sizeable property: the second youngest, in his mid twenties, who had just received the Limonar inheritance. Again, the oldest son, by then in his early thirties, was the poorest. He alternately worked with his father or as a day laborer. The next two sons, in their late twenties, had tiny plots of land insufficient for the needs of their growing families (both had three young children by 1960). In order to make ends meet, both these two sons and their wives entered into various *en companía* arrangements with Aristides to work some of his land in El Palmar. The youngest son was studying in Cali.

We see, then, that the internal relations of the patriarchal unit have a differential effect on the sons, sending some off into the world as wanderers, enabling others to set themselves up as independent economic units, and keeping others tied to the family farm either wholly or as part-time workers. The differential treatment of the offspring may contain some element of personal whim, but the cases of Juan and Aristides illustrate that it may also respond to the logic of the changing demographic status of the household and the changing economic status of the farm over the course of the family/farm life cycle. The older sons, while more likely to want to become independent of patriarchal authority, are also likely to have received less education, because of the importance of their labor input when they were young, and fewer concessions from the patriarch to keep them on the farm as they mature, since by this time their younger siblings would be coming of working age. Thus large family size gives the patriarch a better bargaining position with his older offspring.

This position is no longer as strong with the younger sons as they mature, however, since there is no longer a substitute family labor force waiting in the wings. By this time, too, the patriarch may have more resources with which to bargain. As a result of these changes in the circumstances of the household and the farm unit, the younger sons of Aristides and Juan tended to receive more education, as well as more generous assistance from their fathers when they married. Juan's first three sons were each given small plots of land when they married or reached their early twenties, as well as the *en companía* arrangements, but the younger sons received both better *en companía* arrangements (responsibility for larger and more developed farm areas) and larger plots of land. Finally, in El Palmar it was frequently the youngest son (or sometimes daughter) who remained with the parents as they aged, often becoming the de facto operator of the farm until its subdivision upon the death of the patriarch.

*Economic Environment.* The above analysis suggests that initial differences in the economic status of men entering the family formation stage of their life cycles were related as much to differential treatment by the patriarchs as to differences in the status of those patriarchs. That differential treatment in turn was related to such factors as the personalities of the household members, the position of individual sons in the life cycle of the patriarchal household, and economic conditions affecting the patriarch's investment decisions in land, human capital, and so on (for example, the patriarch's perception of potential returns to the education of some offspring). A key question is whether these economic differences become the basis for economic differentiation—for cumulative differences in economic status as these households mature.

I argued above, in tracing the life histories of that cohort of seventeen men who came of age in El Palmar in the postwar boom period, that this was not the case for this group of men. The economic opportunities of this period acted as an equalizing, rather than disequalizing, force. Those who started out with few resources did not necessarily stay poor, nor did those who began with greater inherited resources necessarily wind up being the better-off households in later years.

That, however, was to change for the cohort coming of age in the 1960s, when the regional and local economic situation deteriorated. Although economic growth had not generated economic differentiation, economic downturn now threatened to do so.

## The 1960s: Commercial Decline and Local Deterioration

In the late 1950s the prices of coffee and pineapples, the key market crops of the smaller farmers, fell. A pineapple glut drove down prices in local markets,[21] while the entry of African coffee into the world market by the late 1950s was beginning to depress international coffee prices. Those coffee prices remained low through the 1960s.[22] The drop in international prices was reflected in a decline in the internal "support price" for high-quality export coffee between 1957 and 1962.[23] Although these prices were raised after the Brazilian frost of 1963,[24] the increases did not compensate for rising inflation. By 1967 internal coffee prices had fallen sharply in relation to the worker and consumer price indices and in relation to basic items purchased by the coffee farmers such as hoes, coffee-picking baskets, beef, rice, maize, and sugar (Zuleta 1975:472–73, tables 8 and 9; Ruiz 1972:2).

In El Palmar, deteriorating market conditions operated in conjunction with the internal dynamic of fragmentation and soil depletion to create a local economic crisis. Relative price declines for the major cash crops in El Palmar raised the minimum farm size necessary to sustain a family, and many small farms were unable to support a household. At the same time that more land was becoming necessary, young families were encountering diminishing inheritances and fewer local opportunities to earn supplemental income. There was increasing competition for fewer jobs.

## *Evidence of Decline: Analysis by Age Cohort*

While the 1960s were generally difficult for the residents of El Palmar, the impact of this period was most severe, and most long lasting, for those households that were just starting out. We can see clearly the effects of the deteriorating economic conditions by com-

21. According to some of the farmers in El Palmar, cultivation of pineapples had been encouraged by plans to build a pineapple-processing factory near Cali. When this venture was abandoned, the bottom fell out of the pineapple market.

22. The average price received per sack of Colombian coffee exported between 1951 and 1957 was U.S. $81.63, compared to an average of U.S. $57.21 from 1958 to 1969 (calculated from Zuleta 1975:440, table 2).

23. The "support price" is the price paid to domestic growers by the Coffee Federation for unroasted (*pergamino*) coffee. This organization had grown steadily in size and power since the 1930s. By the 1960s, according to Palacios (1980:250), it was purchasing over 70 percent of the national coffee harvest.

24. According to Zuleta (1975:500), the ability of the Coffee Federation to raise nominal internal prices in the face of declining external prices was due at least in part to the steady devaluation of the coffee exchange rate.

paring the group of seventeen households that started out in the late 1950s and early 1960s (men between the ages of 40 and 49 in 1978–79)[25] to those households considered above who started out in the prosperous period of the late 1940s and early 1950s (men aged 50–60 in 1978–79). The difference in farm size between the 50–60 age cohort and the 40–49 age cohort is striking: in 1978–79 mean farm size for the seventeen men between the ages of 50 and 60 was 19.2 acres, while for the seventeen men between the ages of 40 and 49 it was only 5.6 acres. While 41 percent of the fifties age cohort had achieved the status of "medium" farmers, and 70.5 percent had acquired more than a "small" farm, only 17.6 percent of the forties age cohort had acquired a "medium" farm, and only 23.5 percent had acquired more than a "small" farm.

This difference was due in part to differences in inheritance size. Inheritances tended to be smaller for the younger cohort than for the older one. The process of fragmentation had proceeded to such a point that only four of the seventeen younger descendants inherited (including early purchase of inheritance) or will inherit five or more acres of land in El Palmar, as compared to ten of the seventeen older descendants. In addition, the economic conditions of the 1960s compounded the problem of small inheritances. Even households who had inherited land found it more difficult to augment their inheritances through land purchases in the 1960s. The largest holding of this group was only twenty-four acres, compared to ninety acres for the cohort in their fifties.

The different experience of these two groups of households was reflected in their economic status in 1978–79: the younger cohort was finding it more difficult to meet the needs of their families than the older cohort. Using a measure of household subsistence food needs based on the age and sex composition of each household, it was found that the ratio of household income to this basic subsistence level was 1.5 on average for the 40–49 age group, and 2.5 for the 50–60 age group.[26] Given the fact that this measure included

25. The average estimated age at marriage for these men was twenty-three. They were men who were children of earlier settlers, or who had arrived in El Palmar as young men in the 1940s or early 1950s. I am not including here two men in this age group who arrived in the 1960s and 1970s.

26. The minimum income necessary in El Palmar to obtain a nutritionally adequate diet in 1978–79 was calculated for each household based on the age and sex

only basic food requirements, with no allowance for other necessities, such as clothing, medical care, candles, fuel, tools, and household utensils, let alone education, home maintenance or improvement, or entertainment, it is clear that the economic circumstances of the younger group were, on the whole, dire.

## *Evidence of Decline: Case Studies*

The case of Carlos Cabrera, oldest son of Lubino, illustrates the difficulties these young households encountered in the late 1950s and the 1960s.[27] Carlos was in his teens when his family moved to El Palmar from El Salado. In the 1950s he worked as a day laborer for his uncle Agapito. When he was twenty-nine, Carlos met Excelinia Hernandez, then sixteen years old, and they married in 1961. Nine months after their marriage, Agapito gave them coffee to work *en companía* for a year. Then, beginning in 1962, his uncle rented Carlos six acres of land for one thousand pesos a year. It had coffee, and Carlos planted pineapples, maize, beans, and household crops. But Carlos had no family labor yet and had to pay hired labor as well as the rent. The prices for his products in the depressed conditions of the 1960s did not permit him to save any money.

In 1968 his uncle sold the farm, and Carlos lost his rented land.[28] By then he and Excelinia had four young children. Carlos went to examine the Caquetá region, a part of the low-lying southeastern plains of Colombia, where colonization was being encouraged by

---

composition of its members, using FAO-recommended nutritional levels by age and sex, the INCAP/ICNND tables of the nutritional content of Latin American foodstuffs, 1978–79 prices of foodstuffs in El Palmar, and observation of the local methods of preparing the foods. The recipes, prices, nutritional calculations, and cost calculations are presented in Reinhardt 1981: Appendix B. The necessary income for each household was then compared to each household's actual income, where actual income was calculated as: income from farm sales in 1978–79, minus farm expenses, plus the imputed value of farm produce consumed by the family (calculated at market value), plus all income from nonfarm sources.

27. Case study information in this section was obtained from household interviews and from oral histories carried out in 1978–79 and 1983–84.

28. Carlos' father, Lubino, however, was able to buy fifteen acres from his brother Agapito. Lubino, at the age of seventy-five, had finally been able to put together a medium-sized farm; but that was too little for his eight children, and too late for his oldest son, Carlos.

the government. He found the area too hot and isolated, and returned to El Palmar to work as an agricultural day laborer. Soon after the birth of another child, in 1969, he and his family moved to El Treinta, where they had gotten a job caring for a *casa de veranéo* (a "weekend farm" owned by an urban family). Carlos, with no land and a growing young family, had been unable to save any money in the difficult years of the 1960s to establish himself in El Palmar. He finally chose to leave the community.[29]

Other descendants in this age group found themselves in similar straits. Luis Mondragón, for example, was a son of one of the original settlers of El Palmar, Pedro Mondragón. Pedro had acquired over fifty acres of land in El Palmar, which he worked with the help of his ten surviving children. Luis farmed some of his father's land *en companía* for many years. He married in 1957, when he was in his late teens. He and his wife had three children in quick succession. In the mid 1960s Pedro told Luis that he should either buy the land he had been farming, or Pedro would pay him for the coffee he had sown. As Luis did not have the money to buy the land, his father paid him four thousand pesos for his coffee trees. Unable to buy land in El Palmar, Luis and his family left the area and spent the next ten years working in other parts of the country. Despite the size of his father's holdings, Luis, with his young family, was unable to establish himself in El Palmar in the 1960s.

Those who remained fared no better than those who left. Luis Murcia arrived in El Palmar as a young boy in 1948. His parents had fled Tolima because of rural violence ("La Violencia," to be discussed below). In 1957 he married Carmen Rosa Enríquez, the daughter of an El Palmar farmer with twenty-three acres of land. In a few years they had two children, and in 1963 a third. But Luis was unable to acquire land. He worked as a day laborer, until finally he and his family got a job as resident managers of a large absentee-owned dairy farm in El Palmar.

29. Emigration was also increasingly becoming an important strategy for the single younger brothers of this cohort. All three of Carlos Cabrera's younger brothers, for example, left El Palmar in the 1960s. The middle brother of these three, Leonardo, left in the early 1960s when he was in his early teens. He worked in the coffee harvests in Quindío for several years, and then in the cotton harvest in Valledupar. Bolívar also went to Quindío in 1965, when he was twenty-one, bringing his fourteen-year-old brother Favio with him. Favio returned after a few months, while Bolívar stayed another year.

Claver Quintero was born in the same year as Luis Murcia, 1936. When he was eighteen he married Gabriela Diaz, the daughter of one of the original settler families of La Virgen. Her father had sixty acres of land, but, as with Luis and Carmen Rosa, this was of little benefit to the young couple. In 1958 Gabriela, one of twelve children, received a 1.5 acre inheritance. For the next ten years, as the family grew (by 1965 they had five children), they struggled to make ends meet. In 1968 they were finally able to buy another 1.5 acres of land, still woefully inadequate for their needs.

## *Economic Differentiation in the 1960s*

By the 1960s, then, the conditions that had sent the original settlers on their journeys to El Palmar had been recreated in this new community. We have traced the roots of the tendency towards fragmentation and decline back to the internal relations of the patriarchal unit and the labor requirements of the family farm. This internal process generated differences in the economic status of the offspring, as we have seen. Yet in prosperous times, as in the 1940s and 1950s, those differences could be countered, and some households were independently able to build up their holdings. At the same time, by switching to more intensive and high-value crops, farmers with small holdings could maintain their income. While family background accounted for some differences in economic status, the more important factors were the nature of each man's relationship with his father, personality, and luck (number of children, ratio of male to female children, health, weather, etc.).

When the market for the small-farm crops declined, however, family background emerged as a key determinant of economic position, and true economic differentiation began to be evident. For the younger cohort, upward mobility was very limited and largely determined by family background. For those in this cohort who had no inherited land, the lack of local opportunities to accumulate savings through wage labor, land rental, or other economic activities locked them into their positions. Of eleven households with limited or no inheritance prospects, only one had managed to acquire over five acres by 1970. These difficulties persisted through the 1970s. The experience of the cohort in their forties strikingly reveals the weakness of the Chayanovian model, which fails to explain the behavior of farm households in a situation of increasing population density and overall economic downturn.

## Conclusion: The Status of Family Farming in the 1960s

By the 1960s the storm clouds that were enveloping older highland communities earlier in the century had reached El Palmar. The processes of population growth, fragmentation of holdings, and general economic decline had caught up with the more recently settled mountain communities, and the localized crises of peasant producers in the 1930s had become generalized. We have seen that the farmers of El Palmar adjusted by switching to more labor-intensive high-value crops, and by adopting more diversified livelihood strategies where possible. Those strategies depended on the participation of household members in a variety of off-farm activities that included temporary and even permanent migration away from the community. As the farmers of El Palmar—and similarly pressured farmers in other parts of the country—made these adjustments, they became more vulnerable to the overall economic climate in which they were operating: less able to "tighten their belts," less able to squeeze a little more out of their land, less able to withstand the vicissitudes of the business cycle. Poverty in the community increased in the difficult conditions of the 1960s, as it did throughout the Colombian countryside; one government study estimated that by 1970, almost 65 percent of rural Colombians were living in absolute poverty (INCORA 1970).

Nevertheless, while the diminishing ability of successive generations to earn a livelihood as farmers over the period between the 1930s and the 1960s was creating mounting rural tensions and an unprecedented migration to the cities, peasant producers as a group still played a key role in the Colombian economy as agriculturalists. Their livelihood difficulties were related to the processes I have described, rather than to growing competition from large-scale agriculture. Chapter 5 shows that, despite important changes taking place in the nature of the production process on Colombia's large estates, those estates were not directly displacing peasant producers from their traditional spheres of staple food and coffee production.

The highway through El Palmar (facing toward Cali).

The roadside market in El Palmar. In foreground: cassava in wheelbarrow, blocks of *panela* (unrefined sugar) on sack, bunches of plantains. In background: meat and produce stalls.

A roadside *fritanga* (fried-food stand) in El Palmar.

Back off the highway: a scattered farm settlement. These four farms belong to offspring of the patriarch Juan Álvarez.

Bringing lunch to the workers in the fields.

One of Carlos Cabrera's young daughters picks coffee with her mother and sisters on a neighbor's farm.

While Aristides Benavides' wife, Romelia, still dries their coffee on the church patio (built on land Aristides donated) . . .

. . . his son Arcesio dries his coffee on his new sliding-cover drying roof.

Mixing pesticides for tomatoes and green peppers.

Tending bees.

Watching television is a new pastime in El Palmar.

Desiderio Almeida (right) with crates of tomatoes ready for the market bus; at left is a local shopkeeper.

*Some Other El Palmar Residents*

Aristides Benavides, patriarch of El Palmar.

Lubino and Ursulina Cabrera.

Carlos and Excelinia Cabrera and their daughters; their sons have left home. Coffee and kidney beans dry in the racks.

Carlos' brother Bolívar Cabrera: a young man on the rise.

*Chapter Five*

# Modernizing Colombian Agriculture: The Estates after World War II

While the economic difficulties of peasant producers were becoming more severe and widespread in the postwar period, the estate sector was gradually undergoing a transformation that gave it new life. By the 1950s traditional cattle and food-producing estates in several regions of the country were becoming modern commercial operations. By the 1960s modern technology was finally being introduced into large-scale coffee cultivation. The changes in Colombian traditional estate agriculture were part of a broader transformation taking place in many Latin American countries, the nature and extent of which researchers are just beginning to document.[1] A key task before us is to understand the dynamic of that estate transformation process: its roots, its potential scope, and its overall impact on agrarian structure. This chapter examines these issues in the Colombian case.

## Estate Retrenchment and Growing Agricultural Shortages

Chapter 3 presented the growing difficulties of the estate sector in the 1920s and 1930s as the *hacendados* faced escalating pressure

1. The earliest work demonstrating the extent of estate modernization in Colombia is Kalmanovitz 1974. For Latin America as a whole, see de Janvry 1981 and López Cordovez 1982.

from their labor force and falling product prices after 1929. The agrarian legislation of 1936 calmed the rural unrest, but did little to change agrarian structure or the conditions faced by both small and large producers. In response to the economic and social conditions of the period, some owners of large coffee haciendas parceled out land to tenant workers, while many estate owners withdrew from food production, getting rid of their tenants and converting that land from crops to pasture. Some commentators of the period spoke of the virtual disappearance of the various tenant arrangements (Machado 1977:331–34; Fajardo 1983:53, 60; Hirschman 1963:114).

This behavior had serious consequences at the national level, where the expanding urban-industrial sector depended on a parallel expansion of agricultural production of foodstuffs, exports, and industrial raw materials. Despite tariff protection, production shortfalls caused agricultural imports to increase in the late 1930s (Fajardo 1983:61–62). The outbreak of World War II cut off imports and resulted in increasing agricultural prices in the early 1940s, which the government ineffectively tried to counter with price controls (Machado 1981:68). Pressure was mounting to confront the dual problem raised by the concentration and underutilization of land on the large estates.[2]

Nevertheless, the solution was not clear. One possibility was a direct attack on the internal frontier of the estates through a redistributive land reform. The alignment of political forces by the early 1940s, with conservative elements in the ascendancy, precluded such a radical reform. SAC instead urged a policy of government support for the development of agricultural productivity as a means of stimulating more intensive cultivation of estate land. The beginnings of this approach were already evident in the late 1930s in the initiation of agricultural research programs and the creation of agricultural credit institutions, such as the Fondo Nacional de Ganadería founded in 1939 (Fajardo 1983:63–64). However, these efforts were still minimal. Decree 1157 of 1940 proposed major investments in agricultural research and institutional and infrastructural development, but World War II limited the implementation of these programs (Machado 1981:67).

2. Later studies attest to the continued problem of concentration of landholdings and underutilization of land on the estates. See, for example, Currie 1950; ECLA 1956; and CIDA 1966.

As the agricultural shortages grew more severe in the early 1940s, an attempt was made to reestablish the forms of production that had served the estates so well in the early decades of the century. Law 100, passed in 1944, reinterpreted the provisions of the 1936 legislation in such a way as to encourage a return to sharecropping on the estates. In order to obtain their cooperation, the landed elite were given a five-year continuation of the 1946 deadline for expropriating underutilized land, if at least one-third of such land was let out to tenants. Presumably, small-scale, family-labor-based production on the estates could expand without the institutional and infrastructural support that would be necessary to develop a unique and economically competitive form of large-scale estate production.

This policy, too, proved inadequate. As we have seen, the rapid urban and industrial expansion of the postwar period sharply increased the demand for agricultural products. Domestic food production continued to grow, but that growth was declining by the late 1940s. According to a study by the UN Economic Commission for Latin America, between 1940–44 and 1950–53 foodstuff production in Colombia increased at an average annual rate of 2.9 percent, with a population increase of 2.5 percent; between 1945–49 and 1950–53, however, foodstuff production grew at only 2.1 percent (ECLA 1956:152, table 113).

As inflation accompanied shortages of agricultural products, real wages declined and urban unrest mounted, along with a bipartisan effort to repress it (Mejía 1978:171–72). Rural-urban migration, a safety valve for rural tensions, simply transferred those tensions to the urban areas. The increasingly serious implications of agricultural underutilization for urban and industrial development were not lost on those groups who favored such development. At the same time, rural unrest increased as the growing rural population faced a severe land shortage.

## La Violencia

The social and political tensions that had been building for decades began to erupt more violently in the postwar period. The assassination of Jorge Eliécer Gaitán, the popular Liberal reformer, set off days of rioting in Bogotá. Far more severe was the increasing rural violence, which reached massive proportions by the late 1940s. Under the Conservative governments first of Ospina Pérez (1946–50) and then of Laureano Gómez (1950–53), traditional landowners throughout the country launched campaigns of repression and retribution against those peasants who had been pressing for access to their land.[3] Rural guerrilla groups were organized in defense, often under the auspices of the Liberal party, while the *hacendados* organized bands of loyal peasants to rout out "troublemakers." In many parts of the country, peasant fought against peasant under the banner of one or the other political party.

La Violencia, as these struggles have come to be known, resulted in the deaths, variously estimated, of from 100,000 to 300,000 people between 1946 and 1956, and the uprooting of hundreds of thousands of rural families. The roots of La Violencia are complex and have occupied volumes of analysis.[4] I would suggest that at least three levels of struggle were involved: between workers and employers, in both the urban and the rural sectors; between the Liberal and the Conservative parties over political control of the country's modernization process; and third, related to the second, between urban-industrial interests and the landed elite over control of the internal frontier of the estates.

This argument about the nature of La Violencia differs somewhat from other current interpretations. La Violencia has been of particular interest to political scientists, who have interpreted it primarily as a struggle between the Liberal and Conservative parties for national political power; to sociologists, who have searched for

3. The roots of the political violence go back to the 1930s (and from there to the political struggles of the 1800s). In 1930 the Liberals had won the presidency but were faced with a Conservative majority in the National Assemblies, which they hoped to overturn in the 1932 elections. They decided to concentrate on the voters in regions where the Conservatives had only a slight majority. Their tactics were frequently violent. Since the Liberals had control of the police and the army, the Conservatives organized local guerrilla groups to fight the Liberals. In 1946, with the election of the Conservative Mariano Ospina Perez, the Conservatives faced the reverse situation and used the tactics that had been used earlier by the Liberals. "They applied terror by doses in certain areas of weak or minimal Liberal preponderance . . . they were extremely bloody and provoked the large-scale emigration of the rural population to the cities and to Venezuela" (Gilhodes 1970:415, 424).

4. Perhaps the best works to date on this topic are the two-volume study by Guzmán et al. (1962, 1964) and the recent work by Oquist (1980). See also the discussions in Fals Borda 1969:142–45; Fajardo 1983:75–86; Gilhodes 1970; and Piñeda Giraldo 1960.

its roots in the national character; and to Marxist scholars of various disciplines, who have emphasized the campaigns of terror by *hacendados* against peasant activists and the organization of peasants in self-defense. Two recent works have added new dimensions to these approaches. Paul Oquist (1980) has enriched the political science interpretation by arguing that the violence stemmed from local conditions. Economic status in rural communities had been traditionally affected by one's political affiliation and depended on which party held national power. When political divisions within the parties led to a breakdown of state control over local conditions, the violence erupted. Charles Bergquist (1986), concurring with this perspective, has eloquently shown how the conditions of existence of peasant producers—their constant individual struggles to assure the means of their livelihood, and the importance of political affiliation to their economic success or failure—were conducive, under appropriate national circumstances, to their participation in violent struggle.

These arguments, while correct and persuasively argued, tell only part of the story. By emphasizing the local political and economic struggles, they draw attention away from the broader context of the agrarian problem, which is where the underlying roots of the violence are to be found. Local struggles took place in a context of intense disagreement among and between members of the Liberal and Conservative parties over agrarian policy. These debates were not new; they had been raging since independence and had led to numerous violent confrontations in the nineteenth century. Both parties had supported modernization of the Colombian economy, but disagreed about which policies were appropriate to achieve that common goal (Reinhardt 1986).[5] The Conservative position was founded on the need to preserve traditional mechanisms of labor control while developing commercial agriculture on the estates. The Liberals, by contrast, increasingly favored the transformation of labor relations in agriculture as a necessary element in commercial agricultural development.

The difference between these two positions can be clearly seen

5. Both parties continued to support the principle of economic modernization in the twentieth century. The Conservative Laureano Gomez attempted to create a favorable climate for industrial development in the early 1950s (cf. Mejía 1978: 175–79; Fajardo 1983: 74).

in the attempt to eliminate tenancy under the 1936 law and the attempt to reestablish it under the 1944 law, in both cases as the basis for the continued expansion of agricultural production on the estates. The struggle between the leaders of the Liberal and Conservative parties was not simply over political power per se, but also over political control of the nature of the country's modernization process and, at its core, over the nature of production on the large estates. These fundamental policy differences were intertwined at the local level with the struggle over access to land—a struggle that, as Bergquist shows, occurred as much among peasant producers as between the peasantry and the landed elite (of both parties).

Certainly other Latin American countries were caught up in similar policy debates. That these debates took such a violent form in Colombia, and the timing of that violence, needs to be explained, and the analyses of the authors cited above help to clarify these phenomena. The nature and intensity of the struggles in Colombia were certainly related to the Colombian political process, as Oquist demonstrates, and to the political, economic, and social conditions of Colombian peasants, as Bergquist shows. In focusing on these particular questions, however, we should not lose sight of the underlying agrarian problem.

The paramount features of the local struggles that engulfed Colombia in this period varied over the course of the violence and from region to region. Although Valle was hard hit, much of the Dagua region appears to have escaped the upheaval. Families fleeing the violence elsewhere, particularly in Tolima, arrived in El Palmar. The relatively egalitarian nature of the community, its recent settlement, and the lack of traditional estates in the immediate vicinity (although there are some in the Dagua region) help to explain the area's comparative calmness. Although there were political differences in El Palmar and its neighboring communities, there were no clear class differences, and relations between households over access to land were not yet as conflictive as in other regions. The relative tranquillity of the Dagua region suggests the significance of economic, rather than political, forces in precipitating unrest.

A resolution of the underlying policy differences ultimately could not be achieved on the battlefield, as the balance of power between

Liberals and Conservatives, and between the urban-industrial and traditional landed interests, was fairly even. Although the country had experienced three decades of rapid urban and industrial growth, the landed elite were well entrenched (Machado 1981:71–72), and the presidency had swung between the two parties. While the battle dragged on, aspects of the class struggle began to intensify both in the countryside and in the industrial sector. By the 1950s the guerrilla groups the political leadership had been using were escaping from their control, and while some of those groups turned to banditry, the politicization of others gave the rural conflicts in some regions the character of class-based struggles over land (Gilhodes 1970:431).[6] The leadership of both parties began to search for a compromise policy that could resolve the underlying problem of inadequate agricultural output in a manner acceptable to as broad a segment of the elite as possible.

## Imports and Import-Substitution: Modernizing Colombia's Agricultural Estates

The favorable international context of the postwar period provided the means for addressing the problem of agricultural shortages with a two-pronged approach that would leave the agrarian structure unchallenged. First, in a return to the strategy of the 1920s, which was once again feasible, the foreign exchange earned by coffee exports would be used in part to finance imports of foodstuffs and industrial raw materials. These imports, particularly wheat, oils, and milk, grew steadily in the early 1950s (Machado 1981:76–77; ECLA 1956:158–59).[7] Second, the program of support for estate modernization, proposed but barely implemented in the late 1930s and early 1940s, would now become the key policy instrument to deal with the problem of agricultural production.

Although the policy of leaving the agrarian structure intact and

6. See also the discussion in Fajardo 1983:80–83, in which the role of the Communist party is emphasized.

7. This corresponded with the needs of Colombia's major trading partner, the United States, which had been facing growing agricultural surpluses after World War II and soon became, under the 1954 Agricultural Trade and Development Act (Public Law 480), a major supplier of foodstuffs to Colombia.

developing the estates was not a marked departure from earlier policy, its institutionalization as the principal agrarian policy for the country was achieved only after a decade (1946 to 1957) of political and social turmoil. Opposition came both from traditional landowners, who were opposed to rapid change, and from some modernizers and social reformers, who feared that this policy would leave the basic agrarian problem unresolved. But support came from many sources. Not only urban-industrial groups but even many segments within SAC favored this approach. In addition, major impetus came from groups outside the country. Most influential was the World Bank, which in 1950 conducted an IBRD (International Bank for Reconstruction and Development) mission to Colombia headed by Lauchlin Currie. The findings were highly critical of land-use patterns (although not land-tenure patterns), and recommended the use of tax policy to force more intensive land use on the estates (Currie 1950). A later report (IBRD 1956) reached essentially the same conclusion.

The gradual strengthening of the developmentalist policy can be traced through the turbulent decade of the 1950s, despite the frequent political changes of the period (the overthrow of Gómez in 1953 by Rojas Pinilla, Pinilla's overthrow in 1957 by a military junta, and the establishment in 1958 of a National Front government that involved the sharing of political power by the Liberal and Conservative parties). In 1949 the Conservative government of Ospina Pérez had already signed an agreement with the Rockefeller Foundation in the United States for the development of agricultural research in Colombia. This led to the establishment of several research stations in the mid 1950s and to the creation of the Agricultural Research Department (Departamento de Investigaciones Agropecuarias) in 1955. In 1953 an extension program was established with U.S. assistance (Machado 1981:77). Financial institutions serving the agricultural sector were created or expanded: the Banco Cafetero and Banco Ganadero were created in the mid 1950s and the government agricultural bank (Caja Agraria) was given large increases in financing, from $13.4 million in 1945 to $45.2 million in 1950 and to $142.7 million in 1960 (Machado 1981:76). Between 1950 and 1960 the proportion of agricultural production financed by government and private credit rose from 6.4 percent to 16 percent (Vallejo Murillo and Valencia 1973), with credit going predominantly to medium and large holdings.

While incentives to agricultural modernization were gradually being established, the "stick" had not yet been abandoned. It was, however, no longer the threat of expropriation. In 1953 Pinilla tried to establish a tax system penalizing inefficient land use, along the lines of the Currie mission recommendations. In 1957 the military junta reiterated this policy with Decree 290, which established differential land taxes and reestablished the Instituto Geográfico Agustín Codazzi to carry out tax surveys and land valuations. These tax policies were adamantly fought by members of SAC and ultimately were not implemented (Machado 1981:74–75). At the same time, Decree 290 contained major tax incentives for agricultural investment, an aspect of the measure that was carried out (Hirschman 1963:126–29).

The use of economic incentives to encourage estate modernization continued into the 1960s. Government policies included price supports for some crops, import subsidies for necessary inputs, continued development of infrastructure, technical assistance, subsidized credit, preferential exchange rates, and tax incentives.[8] Barley production was encouraged by Colombian breweries, which organized a service of technical assistance, distribution of seed and fertilizer, and machinery rental (CIDA 1966: 194). International organizations continued to support the modernization of Colombian agriculture.[9] The modernization of the estates was also aided by the growing rural labor force and declining real wages of the 1948–58 period (Kalmanovitz 1974; Fajardo 1983:72), which resulted from population growth and the rural dislocations of La Violencia.

These incentives, in the context of the rapidly diminishing possi-

8. The government at times set the prices of certain crops above the world market price to encourage domestic production. This was particularly true for sugar, rice, and cotton. Control over the production and distribution of improved seeds was vested in the government agency ICA (Instituto Colombiano Agropecuario), established in 1963. A series of national and decentralized organizations were established to further agricultural development, including IDEMA (Instituto de Mercadeo Agropecuario), VECOL (Empresa Colombiana de Productos Veterinarios), regional development corporations, and organizations to encourage the development of specific crops.

9. The agricultural research institute CIAT (Centro Internacional de Agricultura Tropical) was founded in Valle with funding from the U.S. Agency for International Development (USAID), the Rockefeller Foundation, the Ford Foundation, the IBRD, the International Development Bank, and a host of other international groups. CIAT's research in the 1960s principally benefited Colombia's estate sector.

bilities for maintaining traditional relations of production in the turbulent countryside, induced many members of the landed elite to take up commercial crop production or to rent out their land to those who would. Nevertheless, the transition was not without opposition and struggle.

### *An Example: The CVC in Valle del Cauca*

These developments took place with particular rapidity and force in the Cauca River Valley, site of some of the most fertile land in Colombia. The valley had undergone rapid urban-industrial development in the 1930s and early 1940s: by 1945 Valle was the third department of Colombia in terms of value of industrial output (Ospina Vásquez 1955:502, table 4). Despite these economic changes, the valley estates had continued in their traditional land-use patterns. In the early 1950s, 56 percent of the flat valley lands were under pasture and only 27 percent under crop cultivation (Lloyd-Jones 1961:121). Much of the crop land was cultivated in basic foodstuffs by tenant smallholders, under various tenure arrangements with the large landowners. The only important estate crop was sugar cane, used to make refined sugar in the mills (*ingenios*) of the valley.

Although the rich valley lands had tremendous agricultural potential, the lack of flood control or irrigation facilities made crop cultivation risky. The cost of flood control and reclamation projects was prohibitively high for the landowners (Lloyd-Jones 1961:120). In order to carry out the necessary transformation of valley agriculture, the urban-industrial leaders of the region created the Autonomous Regional Corporation of the Cauca Valley (CVC) in the 1950s. The CVC "was the answer to intense needs for regional development which had been widely felt for many years," according to Antonio and Jeanne Posada (1966:59) in their chronicle of this agency. The Posadas describe the history of the formation of the CVC as follows:

> For years the government and leading citizens of the Department of Valle del Cauca had sought means of quickening and spreading economic growth in their region. . . . in 1945, Dr. Ciro Molina Garcés, the Secretary of Agriculture in Valle, contracted with a New York firm for an engineering study on improved use of the water resources of Valle. . . . Upon

the suggestion of the Cali branch of the Colombian Manufacturers Association (ANDI), the Government of Valle in June 1953 formed a Departmental Planning Committee for assistance and advice. . . . In early 1954, Mr. David E. Lilienthal, former Chairmen [sic] of the Tennessee Valley Authority in the United States, visited Colombia at the invitation of the President of the Republic, and in June 1954, submitted a report to the President. The report reflected the experience of the Tennessee Valley and the T.V.A. (P. 59)

The CVC was created as an autonomous agency with its own source of revenue, a surtax on Valle real estate.[10] In addition, it was given the right of eminent domain. This history suggests that the CVC was formed by the urban-industrial bourgeoisie in order to force "modernization" on the landed elite of the valley, with tax and domain authority to carry out the restructuring. Although some industrialists and merchants themselves had agricultural interests, the idea of the CVC was not enthusiastically received by many valley landowners. A fierce struggle ensued, with these landowners initially fighting the forced tax.[11] In the end a compromise was reached: the CVC carried out extensive reclamation, irrigation, hydroelectric, and other infrastructural projects, which were financed in large part by national and international capital rather than land taxes,[12] and the benefits of these programs were received

10. "The basic financial resources of the CVC initially depended upon a 4 per 1000 real estate surtax levied in the Department of Valle on all properties the owners of which had a liquid patrimony of over Ps. $50,000; and today upon a similar real estate surtax which is, however, only at a rate of 3 per 1000 and affects only owners of real estate whose patrimony exceeds Ps. $100,000" (Posada and Posada 1966:62).

11. Discussions of this struggle are to be found in Bonilla 1967:22; Asociación Colombiana de Ganaderos 1958; and Hirschman 1963:136–37. Bonilla (1967:39, 47) characterizes the resistance of the landowners as "incomprehension when confronted with technology" and "resistance to innovation," but it would seem more likely that this was a clearly recognized power struggle between the landed elite and the urban-industrial elite. The landowners were fighting not only the imposition of the surtax in the 1960s, but also the possibility of losing their land. The CVC at that time began to carry out major projects in conjunction with INCORA, the Colombian Agrarian Reform Agency (the formation of this agency is discussed below). Under the agrarian reform statutes, such improved land was in danger of being expropriated.

12. For example, the Inter-American Development Bank assumed the financing of the Roldanillo-Unión-Trujillo Irrigation District (Bonilla 1967), and the Alliance for Progress provided loans for irrigation projects in the valley (Economist Intelligence Unit 1962–73).

by the landed elite, who retained ownership of their vastly improved lands.[13]

A major transformation of land use took place in Valle, as both cattle land and traditional crop land were converted to sugar cane, irrigated rice, cotton, sorghum, soybeans, sesame, and corn for feed. The number of cattle fell from 818,000 in 1954 to 440,000 only three years later (Asociación Colombiana de Ganaderos 1958), while some 6,400 hectares of maize and beans were turned over to cotton alone between 1954 and 1959 (Mejía 1971:269). By the 1970s only 25–33 percent of the valley land was still being used for cattle grazing,[14] while most of the rest was under crops grown on large-scale units using the latest in imported agricultural technology (IFA 1967; CVC 1978). In 1974 Valle produced 21 percent of the nation's cotton, 26.5 percent of the sorghum, 100 percent of the soy, and 55 percent of the sugar cane (CVC 1975:29, 32, tables 17, 19).

## *Effects of the Strategy: Commercial Crop Production*

Valle represented the leading edge of agricultural modernization in Colombia, but similar transformations were taking place in other fertile river valleys and the coastal lowlands. Production of cotton, irrigated rice, sesame, sorghum, barley, sugar cane, and soybeans increased at an average annual rate of 8.2 percent between 1950 and 1972, with acreage under these crops expanding at 5.2 percent per year (Vallejo Murillo n.d.). The value of these crops increased steadily, from approximately 380 million pesos in 1950 (8.9 percent of the value of Colombian agricultural production) to approximately 1,600 million pesos in 1966, accounting for over 30 percent of the value of Colombian agricultural production through the 1960s (USDA 1970:45, 47; Kalmanovitz 1974:99).[15] Fertilizer use in Colombia increased at an average annual rate of 6 percent between 1950 and 1967, and imports of agricultural machinery, par-

13. This was the case except for small areas purchased by INCORA for redistribution, for which the original owners were handsomely indemnified (Bonilla 1967).

14. This information is based on interviews with CVC officials.

15. The most rapid growth, 15 percent per year, occurred in the period from 1955 to 1959 (Fajardo 1983:72).

ticularly tractors, rose considerably in the 1950s (USDA 1970:65, 74–75).[16] As production of these crops accelerated, parastatal producer organizations were formed. These were organizations that operated "in close cooperation with the government" (USDA 1970:46) to control production and marketing. Cotton production, for example, was stimulated by the Cotton Development Institute and later by the National Cotton Growers Federation (CIDA 1966:194; Arango Londoño 1972:58–59). Similar organizations were created for each of the estate crops.

## *Effects of the Strategy: Foodstuffs and Family Farming*

The modernization of estate production, as we saw in chapter 1, has been argued on both theoretical and empirical grounds to lead to the destruction of peasant production, as the more efficient capitalist units outcompete the peasant producers in both product and land markets. This classic Marxist argument has been developed most forcefully in the Colombian case by Kalmanovitz (1974) in the first major study of the developing structure of Colombian agriculture between 1950 and 1970. Kalmanovitz defines a number of categories of agricultural production in Colombia, including those crops grown primarily on large modern estates ("commercial" crops) and those crops grown primarily by peasant producers ("traditional" crops).[17] He concludes that the expansion of commercial agriculture occurred at the expense of traditional foodstuff production, signaling the gradual displacement of smallholders.

As I argued in chapter 1, however, the evidence on this point is mixed. During the period between 1957 and 1962, in which the Kalmanovitz data show the greatest growth in the value of the commercial crops (from 18 percent to 32 percent of total agricultural value), traditional crops declined only slightly in relative value

16. By 1956 there were 16,493 tractors in the country, compared with 3,821 in 1938 (Fals Borda 1969:125).

17. Commercial agriculture in this study included cotton, irrigated rice, sorghum, sesame, barley, soy, and sugar cane for refined sugar; while traditional agricultural production included plantains, cassava, *panela*, and beans. Kalmanovitz also identifies three other categories of agricultural production: plantation agriculture (bananas and cacao); mixed agriculture, or those crops grown on both large and small holdings (corn, potatoes, leaf tobacco, and wheat); and coffee.

(from 36.4 percent to 33 percent) and increased in absolute value (Kalmanovitz 1974:99). Furthermore, there is some evidence that the number of units under ten hectares and the amount of land controlled by them may have expanded during this decade.[18] In the decade of the 1960s, traditional crops maintained a fairly constant proportion of the total value of agricultural output, while continuing to increase in absolute value, and commercial crops dropped slightly in relative value (from 32 percent in 1962 to 30 percent in 1972) (Kalmanovitz 1974:99). Another government study found that agricultural units under ten hectares accounted for 63.2 percent of domestic foodstuff value added in 1973 (Ortega 1982:85).

While peasant production was continuing to expand, that expansion was difficult. The World Bank mission had already found serious problems of fragmentation and erosion in many regions of peasant production by the late 1940s,[19] with commercial fertilizer in use by smallholders only for potato cultivation. As a result of the deteriorating quality of their shrinking land base, in the context of a scarcity of all production factors other than labor, the overall rate of growth for food production slowed, although the rates for individual crops varied considerably. Between 1950 and 1969 the average annual growth rate of plantain production was 3.2 percent, a high rate probably due to its cultivation as a companion crop to coffee. Potatoes also showed a solid growth rate of 3.2 percent. This was one of the most strongly commercialized peasant crops and was the first to benefit from the use of commercial fertilizer. Other basic foodstuffs had a more dismal record over this 20-year period: beans 1.2 percent per year growth, cassava .9 percent, maize .8 percent, and *panela* .3 percent; while population growth rates surpassed 3 percent annually after 1960 (World Bank 1972:266, table 14–1; IMF 1964–74). With only two exceptions (potatoes and *panela*), the percentage increase in output was greater than the percentage increase in land devoted to a given crop (Velez 1979:645), evidence that by the 1950s land was being used more intensively.[20]

18. This is based on a comparison of the Colombian 1950 sample agricultural survey and the 1960 agricultural census. Since the 1950 data were based on samples rather than a full census for many regions of the country, this comparison is not conclusive.

19. These problems were found in the departments of Huila, Magdalena, Boyacá, Santander, Pasto, and Nariño (Currie 1950). See also Adams and Schulman 1967 and Haney 1969.

20. The figures on smallholding production are notoriously unreliable and

Rising real prices of basic foodstuffs by the 1960s (USDA 1970: 38, 40–41) reflected adjustments in the peasant sector. As household income from farming declined due to shrinking average farm size, more members of peasant households turned to migration. In the Dagua region, migration reduced population growth despite the large family size of the rural households. Between 1938 and 1964, the population of the *municipio* grew at the more modest—although still high—average annual rate of 3.1 percent.[21] The sons and daughters of El Palmar joined the ranks of the migrants.

We can get some idea of this migration process by looking at the offspring of the 105 resident households interviewed in El Palmar in 1978–79.[22] As table 5.1 shows, 18 sons and daughters of these households permanently left the community before 1958, out of a total of 95 sons and daughters of "migration age" (over fourteen) in that period. That left 77 offspring of migration age in 1958. Another 94 members of these households reached migration age between 1958 and 1967, for a total migration-age population of 171. In those ten years, 41 of these sons and daughters permanently left El Palmar, while 130 remained. Between 1968 and 1978, another 169 sons and daughters of these households reached migration age. Out of the total of 299 resident offspring of migration age between 1968 and 1978, 73 migrated. The total number of sons and daughters of these households who had migrated from El Palmar by 1978 was 136, including 4 migrants for whom date of migration was unknown. This represented 38 percent of the adolescent and adult offspring of these households.[23]

must be treated with extreme caution. In El Palmar a comparison of the results of the 1977–78 agricultural census (original returns) with the data collected in my interviews revealed considerable underreporting in the census. The extent of foodstuff production, and of acreage devoted to that use, is probably understated in the official statistics for the 1950–72 period. However, there is no reason to suppose that the understatement had been increasing over time—if anything, the opposite may be true. The overall trend suggested by the available data, of an increasing inability of basic foodstuff production to keep pace with population growth in the 1950s and 1960s, is therefore probably correct.

21. The population in 1938 was 14,149, while that of 1964 was 25,508 (see Appendix A).

22. These data, obtained from the household interviews, include only the offspring still living in 1978–79.

23. It should be noted that some of the heads of household and their spouses were themselves offspring of other households in the study, and were included in these calculations.

TABLE 5.1. *Permanent Migration from El Palmar: Sons and Daughters of 105 Resident Households*

| | *(1) Off-spring Reaching Age 15 in Period* | *(2) Off-spring in El Palmar Age 15 or Older in Period* | *(3) Of (2), Number Migrating during Period* | *(4) Of (2), Percentage Migrating during Period* | *(5) Offspring 15 or Older at End of Period* |
|---|---|---|---|---|---|
| Before 1958 | 95 | 95 | 18 | 18.9 | 77 |
| 1958–67 | 94 | 171 | 41 | 24.0 | 130 |
| 1968–78 | 169 | 299 | 73 | 24.4 | 226 |
| Total[a] | 358 | 358 | 136 | 38.0[b] | 222 |

*Source:* Household interviews by author.
[a] Including four offspring for whom dates of migration are unknown.
[b] Percentage of all sons and daughters aged 15 or older who have permanently migrated from El Palmar.

Table 5.2 presents information on the reasons for migration. Of the total of 73 daughters who left El Palmar, 48 (65.8 percent) left because they married. Of the remainder, most (20 out of 73, or 27.4 percent) left to work in Cali or other cities. Of the 63 sons who left El Palmar, the majority left to look for work or land. Twenty-six, or 41.3 percent, went to look for work in the city (16 between 1968 and 1978). Twenty-three, or 36.5 percent, went to look for land or work in the countryside. Of these 23 men, only 4 had acquired land by 1978–79.[24] These data show clearly that migration from peasant communities could no longer be primarily a search for new land to settle. Migrants were now highly dependent on employment in the coffee regions, on the commercial estates, and in the cities.[25]

These migrations had the important effect of improving the

24. Including Alfredo Benavides, whose father Aristides had, as we have seen, bought him a farm in El Limoner.

25. These population movements in the 1950s reflected not only the expanding employment opportunities in those areas and rural overcrowding, but also the dislocation of La Violencia. After the late 1950s, however, this was no longer an important factor.

TABLE 5.2. *Reasons for Permanent Migration from El Palmar: Sons and Daughters of 105 Resident Households*

| | *Marriage* | *Military Service* | *Study* | *Work in City* | *Land Elsewhere or Work in Countryside* | *Reason Unknown* | *Total* |
|---|---|---|---|---|---|---|---|
| Daughters: | | | | | | | |
| Before 1958 | 8 | 0 | 1 | 1 | 0 | 0 | 10 |
| 1958–67 | 16 | 0 | 1 | 6 | 0 | 0 | 23 |
| 1968–78 | 22 | 0 | 2 | 13 | 1 | 0 | 38 |
| Date unknown | 2 | 0 | 0 | 0 | 0 | 0 | 2 |
| Subtotal | 48 (65.8%) | 0 | 4 (5.5%) | 20 (27.4%) | 1 (1.4%) | 0 | 73 (100%) |
| Sons: | | | | | | | |
| Before 1958 | 1 | 0 | 0 | 4 | 2 | 1 | 8 |
| 1958–67 | 0 | 0 | 1 | 6 | 11 | 0 | 18 |
| 1968–78 | 4 | 3 | 4 | 16 | 8 | 0 | 35 |
| Date unknown | 0 | 0 | 0 | 0 | 2 | 0 | 2 |
| Subtotal | 5 (7.9%) | 3 (4.8%) | 5 (7.9%) | 26 (41.3%) | 23 (36.5%) | 1 (1.6%) | 63 (100%) |
| Total | 53 (39.0%) | 3 (2.2%) | 9 (6.6%) | 46 (33.8%) | 24 (17.6%) | 1 (0.7%) | 136 (100%) |

*Source:* Household interviews by author.

income of the patriarchal farm households, through a variety of mechanisms. First, some of the income of migrants was sent home. Second, the migrants swelled the ranks of those dependent on purchased food, raising the demand for farm products. Third, the shifting of labor out of farm households may have reduced farm production to below what it would otherwise have been. The second and third effects combined to raise prices of foodstuffs, thereby equalizing to some extent the return to farm labor and the market return to off-farm labor.

As I suggested in chapter 1, this is the behavior of agricultural output and prices that we would expect to see in a developing market economy with foodstuff production dominated by peasant producers. In the context of expanding off-farm employment opportunities and a limited peasant land base, livelihood strategies of peasant households became more complex. I believe that this more accurately explains the observed slowdown in the rate of growth of peasant production in this period than does the theory that peasant production disintegrated because of competition with the modernizing estates. The extent of competition in product and land markets between peasant producers and the modern estates was much less than has been suggested.

First, competition for land between the modernizing estates and the independent peasant sector was minimized because of the nature of the new commercial production. The modernization of the estates had finally involved the development of a specific large-scale production process. Estate reliance on the patriarchal household form of production was replaced by mechanization and the use of a seasonally variable wage-labor force, to supplement a small permanent wage-labor force. This production process was most easily implemented in the flat lands of the river valleys and coastal lowlands, rather than in mountain terrain, where mechanization and the supervision of a large labor force were difficult. In these valleys and lowlands, the development of the new labor process displaced traditional tenant households from access to estate land. This displacement can be seen clearly in Valle del Cauca, where the percentage of units under sharecropping or mixed tenure forms (often owner/sharecropper) dropped sharply between 1960 and 1970. For units under five hectares (12.5 acres) the combined percentage fell from 18.5 percent to 8.1 percent, while for units between five and ten hectares the combined percentage dropped from 29.4 percent to 15.5 percent (DANE 1973a:56–60).[26] Statistics for the nation as a whole also reveal the displacement of tenant families, as sharecropped units declined from 13.5 percent of all units under ten hectares to only 6.7 percent (DANE 1973a:26, 29, tables 5, 10).

However, the modernizing estates did not compete for land in the mountain regions where most of Colombia's family farms were located. The agricultural census reports for 1960 and 1970 show a slight decrease in the number of units under ten hectares, since these enumerations were based on operating rather than ownership units. This decrease reflects the displacement of tenant farmers on the modernizing estates, rather than the displacement of independent peasant producers. Between 1960 and 1970, the number of owner-operators with fewer than ten hectares, and the acreage owned by them, increased, as was shown in chapter 1. In 1970 these owner-operated smallholding units accounted for 49 percent of all farm units in Colombia, and smallholding units of all types accounted for 73 percent (DANE 1973a:26, 28, 29, tables 6, 7, 9).

Nor did the modernizing estates compete directly with mountain farmers in product markets. Rather, they concentrated on products where technological and market conditions were more favorable than they were for tropical staples: products for export (e.g., refined sugar), for domestic industry (e.g., barley; cotton; sorghum, corn, and soy for feedgrain processing), and for the higher-income domestic market (e.g., refined sugar, vegetable oils). Their choice of products was affected by three sets of factors: the nature of the markets, the technical conditions of production, and government policy.[27] Many of the products selected by Colombia's modernizing estates had the advantage of readily available technology, specifically developed for large-scale production, that could be imported from the industrialized countries. Some had the

26. Taussig (1974) provides an excellent history of the effects of the modernization of Valle agriculture on the tenants and small farmers of the valley.

27. De Janvry (1981:169) highlights the latter two factors, suggesting that "we can expect capitalist production in each country to develop in those crops that are most amenable to economies of scale and to a high productivity of labor because of technological or natural reasons and because of state protection through tariffs and institutional rents."

advantage of a centralized, relatively stable market (e.g., barley for domestic breweries, feedgrains), often dominated by parastatal producer organizations. And most did not have to compete with an established sector of small-scale producers.

Staple food production had none of these advantages: small-scale production was firmly entrenched, with atomistic markets that would be difficult to control, highly unstable prices,[28] and no available technology that could give large-scale production an edge.[29] Furthermore, the importance of these foodstuffs as basic consumer items for the wage-labor force created political pressure to keep their prices low, making it difficult for the government to encourage estate production of these products through price supports.

Rice was the principal exception to the pattern I have described. Although a basic foodstuff already being produced by smallholders under rain-fed conditions, it was also adopted on a large scale by the estates. This adoption was due to two factors: the existence of readily importable irrigated-rice technology and the actions of the Colombian government in the early 1960s to stimulate estate production of this crop. Government policy included reclamation programs to bring land under irrigation, extension of credit and services, the maintenance of a high support price, and assistance in the formation of a rice-growers' association controlled by the irrigated-rice growers. These programs successfully stimulated large-scale commercial rice production. However, the internal market was limited, so that by 1965 increasingly large stocks began to accumulate. Since the IDEMA (government marketing and storage agency) support price was above the world market price, this meant that it had to purchase a significant portion of the national crop, which by 1968 it was unable to do. As a result, the true market price fell considerably below the support price (World Bank 1972:282).

The government had decided to intervene in this case because

28. Price instability was due to the competitive nature of the markets, the perishability of many of the products, and the lack of storage facilities in the face of seasonal fluctuations in production. I observed this extreme price variability in the markets of the Dagua region and Cali.

29. A study of Colombian agriculture by the U.S. Department of Agriculture (USDA 1970:23) concluded that under the existing state of technology, there were few or no economies of scale associated with the production of these tropical staples.

rice was an important urban foodstuff, worker protests over rising food costs were increasing, and rice was being imported to meet domestic production shortfalls while foreign exchange was becoming scarce after 1957. The example of rice illustrates the requirement of the estates, even where improved technology was available, for the assurance of high and stable support prices before they would embark on the commercial production of a basic foodstuff. This necessitated government intervention—but in the context of insufficient effective demand, artificially high prices inevitably generated surpluses and the need for further intervention. Similar programs to stimulate estate production of other basic foodstuffs would have been even riskier than the rice program, since improved technology was not available for those crops. The National Front governments chose not to undertake such ventures.

By the 1960s a new division of production was emerging in Colombian agriculture. Large cattle estates still operated in a number of regions, continuing to rely in many cases on traditional labor relations. But modern estates had replaced the traditional operations on the best lands of the valleys and coastal plains with crops for domestic industry and high-income markets, and with nontraditional exports. The patriarchal households, increasingly concentrated on the mountain slopes, continued to produce traditional staples and coffee. Finally, coffee was also being produced on large farms. This was the principal area of direct competition between large and small producers. As the 1955 coffee census made clear, however, the family farms still predominated. The average amount of land planted to coffee was only 3.3 hectares per coffee farm. Farms with under 10 hectares in coffee accounted for 94 percent of all coffee farms and 63 percent of all coffee production. These were predominantly owner-operated units (Palacios 1980:228–29; Bergquist 1986:301–3).

This emerging pattern of agricultural production did not by itself adversely affect the mountain smallholders. In fact, the modernization of the estates may have benefited them. While the expansion of commercial agriculture did not compete with the mountain producers in product and land markets, it did displace competing foodstuff production on the former tenant smallholdings, and create an expanding wage-labor force wholly dependent on purchased foodstuffs.

## A Model in Trouble in the 1960s

Although the developmentalist policy successfully transformed Colombia's estates, and although that policy did not directly harm the mountain smallholders, it failed to relieve the pressures building in the peasant sector. The overall development model that had been consolidated in Colombia by the late 1950s assumed that industrial development and the modernization of the estates would indirectly resolve the agrarian problems, thereby obviating the need for agrarian reform. The commercial agricultural estates would be expected to provide job opportunities in the countryside. Urban-industrial development would be expected to absorb any surplus rural labor, with the possible assistance of colonization in marginal areas of the country. Foreign exchange earned by agricultural exports could, with the development of import-substitute agriculture and industry, be used to import foodstuffs for the growing wage-labor force.

By the late 1950s the weaknesses of this model were becoming apparent. With rural population expanding and demographic pressure increasing, conflicts between the peasantry and the landed elite continued even after the formation of the National Front and the end of the most intense period of violence. New political factions emerged with bases in the peasant movements (Gilhodes 1970:434–46; Mejía 1971:275–77). Open unemployment and underemployment increased, particularly in the urban areas, as the capital-intensive development of industry and commercial agriculture was unable to absorb the growing numbers of landless workers (Mejía 1978:226). The Cuban Revolution suggested to the Colombian elites the potential outcome of these trends.

At the same time, external conditions were changing as international coffee prices fell after 1957. This price decline was compounded by declining Colombian coffee yields, which became evident in the late 1950s after several decades of stagnation.[30] Smallholding coffee production was suffering from the same difficulties as smallholding foodstuff production. The volume of Colom-

30. I draw this conclusion based on the statistics presented by Palacios 1980: 237, table 34. Decreased yield was probably due to the aging of the coffee plantations and declining soil fertility in the older coffee-producing regions, combined with expansion to more marginal land in many parts of the country.

bian coffee exports, which had expanded steadily over the course of the century, began to stagnate. While Colombian coffee exports averaged 5,615,800 bags per year between 1951 and 1955, they averaged only 5,537,200 bags per year between 1956 and 1960 (calculated from Palacios 1980: 232, table 32). Stagnant coffee exports and falling international coffee prices threatened not only industrial growth (highly dependent on foreign exchange for imported machinery, parts, and raw materials, which the import-substitute agricultural sector was not yet supplying in sufficient quantities)[31] but also the ability to maintain food imports. The growth of gross domestic product slowed in the 1960–70 period to an annual average rate of 5.1 percent (World Bank 1979: 128), while food prices rose.

## *Agrarian Reform*

In response to these pressures, and with the prodding of the U.S. Alliance for Progress program, the National Front coalition passed an agrarian reform law in 1961 that created the agency INCORA (Instituto Colombiano de Reforma Agraria).[32] The history and accomplishments of this agency have been the subject of much debate and fill many volumes (DANE 1970; INCORA 1971; Arrubla 1976; Hirschman 1963; Felstehausen 1971; Kalmanovitz 1978; Fajardo 1983). What is relevant here is that the redistribution of estate land as a means of dealing with rural unrest was adamantly opposed by SAC and was not politically possible.[33] This was soon reflected in official statements by INCORA officials to the effect that there was no problem of unequal distribution of land ownership in Colombia (Mejía 1971: 289–91), despite the evidence of the 1960 agricultural census which showed that 6.8 percent of the farm units in the country operated 75.8 percent of the land in farms (DANE 1973a: 26).

INCORA instead emphasized productivity-increasing infrastructural development, particularly hydroelectric facilities, and elabo-

31. Much of Colombia's industry was involved in the processing or transformation of agricultural products: textiles, soft drinks, and tobacco; and later, vegetable oils and feeds for livestock and poultry.

32. On the formation of INCORA, see Hirschman 1963: 141–44; Mejía 1971: 275–79; and Fajardo 1983: 97–102.

33. Fajardo suggests that this was already clear in the debates over the formulation of the reform law.

rate colonization schemes in low-lying regions of the country like the southeastern province of Caquetá. Because of disease, climate, and the lack of transportation routes, these regions had not yet been occupied. Their colonization required large outlays by INCORA, as a result of which only a minimal number of families (6,700) were settled in the first decade (Mejía 1971:298).[34] The infrastructural projects were also very costly and absorbed the major part of INCORA's budget (Fajardo 1983:103–5). By 1970, despite the emphasis on these projects, only 11,900 hectares had been irrigated by INCORA, and only some 40,000 families had been affected by its land distribution programs (Fajardo 1983:105, 111). Clearly, the INCORA programs did not address the problems of peasant land scarcity and poverty, insufficient production of basic foods, and stagnating coffee output.

## Dawning of a New Approach

Yet equally clearly, those problems needed to be addressed. No longer merely personal or community crises, the problems of the peasant sector threatened to generate a new national crisis in the 1960s. The Colombian Coffee Federation, which had acquired considerable social, economic and political power over the previous decades (Fajado 1980:chaps. 10 and 11), was particularly affected by the internal and international developments. On the one hand, the Federation, which represented elite coffee interests, was attempting to maintain internal support prices high enough to protect producer income. The coffee estates were most dependent on favorable coffee prices, because the dislocations of La Violencia and the labor demand of the modernizing estates had finally forced them to rely increasingly on wage labor. Severe repression of the labor movement had already been necessary in the early 1950s in order to keep wages low enough for large-scale production to remain competitive. While declining coffee productivity also raised

34. Recall that migration to Caquetá had been considered by Carlos Cabrera in the 1960s, but rejected because of the difficult conditions faced by the settlers. It is unclear, furthermore, how many of the colonists remained. In El Palmar there were two families who had participated in the colonization of Caquetá, and who left after only a few years because of the extremely difficult conditions they encountered. A good discussion of these conditions is presented by Ortiz (1984).

the necessary support price, falling international prices made it increasingly difficult to sustain those internal prices.

At the same time, the Federation itself, through the Fondo Nacional de Café, was intervening directly in the purchase, stockpiling, and sale (both internal and external) of coffee, and was increasingly squeezed by its own attempts to protect producer income. By 1961 the Federation was running a deficit on its transactions, which recurred in 1963 and 1964 (Palacios 1980:253, 257, tables 38, 39).

## *Modernizing Colombian Coffee Production*

Although the early activities of the Coffee Federation had emphasized the consolidation and efficiency of coffee-marketing operations, it had also recognized the importance of improving the level of technological development in coffee production. The decidedly secondary emphasis placed on technological issues in the early decades of the Federation was probably due to the rapid expansion of smallholding production in the 1930s and 1940s, without technological assistance, and the resultant steady increases in coffee production. Nevertheless, by the 1950s the Federation was devoting more resources to technical assistance and the development of a fertilizer industry (Palacios 1980:224). The events of the late 1950s and early 1960s confirmed to the Federation the need to revolutionize Colombian coffee technology, as a means of enabling Colombian producers to withstand declining prices stemming from growing international competition. The emerging program to develop coffee technology also included, as an important component, a plan to "rationalize" Colombian coffee production by concentrating and consolidating it on the most suitable soils, while encouraging family farmers to diversify, or to switch completely in the more marginal areas.

This approach was initiated with the Five-Year Program for the Diversification and Development of the Coffee Zones (1963–69), started in 1963 with a pilot project in the major coffee-producing region of Caldas. This program, administered by the Fund for the Diversification and Development of the Coffee Zones (FDDZC), involved both public and private agencies, including the National Coffee Federation, the Banco Cafetero, INCORA, the Caja Agraria, and others, and received technical assistance from CIDA, the Inter-American Committee for Agricultural Development (World Bank 1972:264–67).[35]

By the mid 1960s, the technical division of the Coffee Federation was recommending to producers in the principal coffee regions that they replace traditional coffee varieties with the new high-yielding variety, *caturra* (*C. arábica caturra*), and institute new cultivation practices that would ensure the highest yields with this new variety: closer spacing of the smaller *caturra* trees, the elimination of intercropped foodstuffs, sharp reductions in the use of shade trees, and the use of chemical fertilizer and pesticides. This package of practices has come to be known in Colombia as the "technification" of coffee cultivation. In the more marginal regions, peasant producers were encouraged to replace some or all of their coffee trees with other crops and to technify any remaining coffee production. In order to make the diversification proposals more attractive, the FDDZC implemented an extensive program of technical assistance to these smallholding coffee producers. In addition to instructions on new agricultural activities, such as growing tomatoes and other nontraditional foodstuffs, beekeeping, and pisciculture, the program included information on the use of nontraditional inputs (improved seed, chemical fertilizer, and pesticides) and credit to finance the changes.

## *Modernizing Peasant Agriculture*

For the first time, peasant producers on a large scale had access to productivity-increasing technology that could supplement the traditional inputs of land and labor. At the same time, INCORA projects brought infrastructural improvements such as roads and dams to peasant producers in various parts of the country. By the early 1970s another program was implemented that had as its goal the modernization of peasant production: the Rural Development Projects of the Colombian Agricultural Institute (ICA), which were based on a series of studies of smallholding communities outside of the coffee zones (ICA 1972 and 1973). By 1975 there were twenty-

35. Much of the following information on the FDDZC programs was obtained through observations in El Palmar and through interviews with Coffee Federation extension agents in Dagua and in Nariño. Analyses of these programs include CIDA 1967; IDB 1972; and FEDECAFÉ 1977. Bergquist (1986:371) also provides a succinct summary of the new coffee technology, as does Palacios (1980:235).

two of these projects being carried out in various parts of the country (Rochin and Londoño 1975:2), which were then incorporated into a comprehensive national program for the modernization of noncoffee smallholding production, Programa DRI (Desarrollo Rural Integrado, or "integrated rural development"). Programa DRI was the production arm of a broader nutrition program, PAN (Plan Nacional de Alimentación y Nutrición), which was established in the National Planning Ministry of the government. By 1976 DRI projects were being initiated for a five-year period (1976–81) in eight departments.[36]

Why had the Colombian government, after decades of neglect, now begun to direct resources to peasant producers? First, rural unrest had increased as rural conditions deteriorated in the 1960s (recall that INCORA estimated a rural absolute poverty rate of 65 percent by 1970). In 1968 President Lleras Restrepo formed the rural producers association ANUC (Asociación Nacional de Usuarios Campesinos), as a part of his attempt to deepen the agrarian reform. This organization's activities intensified and channeled the unrest. By the late 1960s land invasions were taking place throughout the country, with more than 2,000 in early 1972 (Kalmanovitz 1978:286).[37]

One means of dealing with the rural unrest was through repression. Under the Conservative President Misael Pastrana (1970–74), attacks against ANUC and other peasant organizations were mounted with increasing frequency by government forces and the landowners whose estates were threatened (Fajardo 1983:120). The policy of commercial agricultural development had entrenched the rural elite more firmly in economic and political power and had ended the possibility of land redistribution (Kalmanovitz 1978:288–89).[38] Yet some means of ameliorating rural conditions were clearly necessary to prevent a return to the rural violence of the 1940s and 1950s.

A second factor leading to the consolidation of the smallholder

36. Thirty-two percent of the funding was provided by $130,000,000 in loans from the Canadian International Development Agency, the World Bank, and the Inter-American Development Bank, with the Colombian government providing the remainder (Latin American Regional Report 1982; Dirección General DRI-PAN 1982b:3–4).

37. According to his figures, by 1973 there were 870 consolidated peasant organizations operating on invaded land around the country.

38. This was formalized in Laws 4 and 5 of 1973, which effectively eliminated

modernization strategy was the changing international and national conception of peasant producers. For decades these producers had been considered to be marginal, tradition-bound, risk-averse, and resistant to integration into the market. By the 1960s that conception was beginning to change. Theodore Schultz (1964 and 1965) was the most vocal early advocate of the position that peasant producers were economically rational and efficient producers.[39] By the early 1970s that position was being widely supported.[40] The first major international experiment with the smallholder modernization strategy was initiated in Mexico in 1967, with the assistance of the Ford and Rockefeller Foundations and the international agronomy center, CIMMYT. The results of this Puebla project suggested the validity of the so-called pro-peasant position: with access to technical assistance, commercial inputs, credit, transportation, and markets, peasant smallholders responded favorably by adopting new technology and by increasing marketed output (de Janvry 1981:234–35).[41]

Colombia's agrarian difficulties, the increasingly positive view of peasant producers, and the experience of Plan Puebla were all bases for the development of the ICA programs. By 1973 the World Bank had formally developed a strategy of rural lending for the modernization of smallholding production, particularly in "integrated" rural development projects (McNamara 1973; World Bank 1975; Yudelman 1976),[42] which encouraged the further development of the Colombian program. The smallholder program initi-

---

agrarian reform, while encouraging large-scale agricultural development (Bejarano 1978:258; Fajardo 1983:122).

39. Another early formulation of the position that peasant agriculture could be a positive force for technological change was that of Owen (1966).

40. In addition to studies inspired by the works of Schultz, the "pro-peasant" (*campesinista*) position was being argued by members of the Wisconsin Land Tenure Center (e.g., Dorner and Kanel 1971), by Johnston and Kilby (1975), and increasingly within Latin America itself. This view was strongly opposed by the advocates of large-scale agricultural development, who considered peasant producers to be inherently inefficient.

41. One difficulty, to which we shall return, was that only farmers with sufficient land were able to take advantage of the technology offered by the program (Fajardo 1983:125).

42. Here mention should also be made of the role played by international agribusiness in the design of the new seed-technology packages, in the hope of capturing the vast inputs market represented by the peasant producers of the world.

ated in 1974 by the administration of Colombian President López Michelsen was put together with the assistance of World Bank and FAO advisers. The formal funding proposal for Programa DRI was sent to these agencies the following year (Fajardo 1983:128).

We are still left with a puzzle: why, given the earlier strategy of modernizing Colombia's agricultural estates, did government policy not focus on increasing basic foodstuff production on those same estates? Why switch to a "pro-peasant" policy? One answer, suggested by the objectives of the López program, was the pressing need by the 1970s to absorb labor in the countryside in order to reduce rural-urban migration (Bejarano 1978:265–66). Although family farming promised to be more labor-intensive than large-scale operations,[43] commercial agriculture had absorbed an increasing number of workers between 1950 and 1971, rising from 4.8 percent of the rural labor force in 1950 to 11.7 percent in 1971 (Kalmanovitz 1978:190, table 5-A).[44] The most important factor leading to the strategy of focusing on peasant food producers was probably not their greater employment potential, but rather an awareness that basic foodstuff production was not attractive to the commercial estates, for the reasons outlined above.

The impact of these various programs on Colombia's peasant producers is currently a subject of intense debate in Colombia. Much attention has been focused on the coffee technification pro-

43. This was certainly the case with traditional producers: many studies document the greater labor input per acre on the peasant holdings as compared to the estates (for Colombia see, for example, Barraclough 1973:180–81; for Latin America in general, see Barraclough and Domike 1966; for a cross-regional perspective, see Dorner and Kanel 1971). It should be noted that the demand for labor on the large commercial estates would far exceed labor demand on the traditional estates, despite the mechanization of the commercial estates. Even so, given the labor-intensive nature of the new technology packages being proposed for the peasant producers, we could expect small-scale commercial production to continue to be more labor-intensive than large-scale commercial production.

44. As Kalmanovitz (1978:184) puts it: "the new and more extensive areas under commercial crops require virtual armies of workers. . . . This army migrates to the coast to collect cotton, passes to Caldas to collect coffee—which continues to be the harvest that requires the most wage labor of all the crops—and stops throughout Tolima, Valle de Cauca and the sabana of Bogotá for the less demanding harvests of rice, soy, sorghum, maize and barley." We must remember, however, that at least some of the expansion of commercial agriculture involved the displacement of tenant farmers, whose disemployment must be subtracted from the employment figures for the estates to determine the net employment effect.

gram, and a number of observers have concluded that the development of modern technology is finally enabling large farms to displace peasant producers from Colombian coffee production. This has been argued in relative terms, by pointing to the clear decline, evident in the 1970 census, in the percentage of coffee farms, of coffee acreage, and of coffee output that was accounted for by peasant producers.[45] This relative decline does not necessarily, however, reflect an absolute decline in peasant coffee production. Chapter 6 shows how patriarchal farms in the marginal coffee area of Dagua selectively adopted from the technification package in order to expand coffee cultivation in the 1970s, a decade of rapidly rising coffee prices.

Programs such as Programa DRI whose explicit goal was to modernize peasant agriculture have also been criticized. Some observers have argued that these programs intensify differentiation among peasant producers, helping the richer peasants to become established commercial farmers and accelerating the conversion of poorer peasants into landless wage laborers (Feder 1976 and 1977; Payer 1980). We can examine this argument, as well, by looking at the experience of the El Palmar farmers in the past several decades. While El Palmar was not a DRI site, the farmers were repeatedly visited by extension agents affiliated with the FDDZC program, PAN, INCORA, and a number of other national and even international programs. Agricultural production in the community was rapidly "technified" and "modernized" in the 1970s. These changes, and their impact on the absolute and relative income positions of the El Palmar households, are discussed in chapter 6.

## Conclusion

The modernization of the estates, while beginning to address the problem of inefficient utilization of Colombia's "internal frontier," did not provide a solution to the multifaceted agrarian problem. The growth of wage-labor employment in estate production could not absorb the growing population of the patriarchal farm sector,

45. See, for example, the figures presented by Palacios (1980:245), who does not, however, conclude that peasant producers are necessarily being displaced, and Fajardo (1983:146), who does.

while the modernization of estate production cut off the possibility of transferring estate land to that sector. By the 1960s, with the closing of the mountain frontier and a general decline in market conditions faced by smallholding producers, emigration had become the principal response to demographic pressure. Migration now took a different form and had national ramifications distinct from those of the nineteenth-century settlement migration. In particular, the growing rural crisis now began to transfer itself to the urban setting, manifesting itself in urban overcrowding and rising domestic food prices. At the same time, the "internal frontier" of the estates invited continued rural tensions.

Other responses were tried and found to be inadequate to deal with the rural crisis, responses such as agrarian reform legislation, colonization, food imports, and rural repression. The attempt to modernize the family-farm economy emerged as another possible solution. It emanated from the particular role that the family-farm economy had come to play in the economic life of the nation. In responding to its own internal pressures as well as to external market opportunities, the family farm had expanded into productive activities that were particularly suited to its resource endowments of abundant labor, but limited land and capital. It had become the principal producer of tropical staples, and an important producer of coffee. At the same time, migrants from this sector fed the labor markets of rural and urban commercial production. It was only logical that a crisis in this sector would send waves throughout the economy, generating necessary responses at the national level. Initiatives to modernize peasant production were then to operate on the family-farm economy in the 1970s along with the direct and indirect effects of rising coffee prices, as shown in chapter 6.

*Chapter Six*

# Modernizing the Family Farm: The Economic Transformation of El Palmar, 1968–83

Farm practices in El Palmar in the 1960s were much what they had been since early settlement days.[1] Although techniques of cultivation varied by crop, they had in common a reliance on inputs available on the farm: seed from previous crops, compost of farm and household materials, and family labor. There were also the use of hired labor, *en companía* arrangements, and some labor exchange. Farm implements were rudimentary: the machete, the hoe, the shovel. Metal components were purchased, with wooden components (e.g., shovel handles) fabricated on-farm. The shoulders of the workers were sometimes supplemented by the use of horses or mules for transport and for milling sugar cane.

Aside from *panela*, processed principally by the larger landowners in the community, the one area that involved some development of farm technology was the processing of coffee. In Colombia, unlike other parts of Latin America, the harvested coffee beans are processed by the farm households. This involves stripping the pulp of the coffee berry from the seed (the coffee bean), and washing and drying the stripped beans. Most farm households in El Palmar had purchased a small hand-depulper. The rest of the process-

1. Information in this chapter is based on formal interviews (see Appendix B) with 105 resident households in El Palmar and observations and discussions during the period from September 1978 to June 1979, and a follow-up study in December 1983–January 1984 in which a sample of households was resurveyed.

ing was carried out using simple devices designed and fabricated by the farm families.[2]

The farmers of El Palmar had never used formal bank credit, although they frequently borrowed from family members, neighbors, and local moneylenders, and purchased goods on credit (*fías*) from shops in El Treinta and Dagua. They did not purchase seed, rarely used chemical fertilizer or other agricultural chemicals, and had not received technical advice or assistance on any crop other than coffee until the operation of the FDDZC program.[3]

In the 1960s FDDZC personnel began to encourage farmers in this "marginal" coffee area to switch to other crops. These efforts bore little fruit during the depressed conditions of the 1960s. Despite falling real coffee prices, the small farmers were unwilling to pull up their coffee trees for what they saw as risky or insufficiently remunerative ventures. While some of the recommended projects involved considerable investment (beekeeping or fish ponds, for example), others were not feasible without good transportation connections to urban markets (tomatoes and other perishable fruits and vegetables). The road connecting the Dagua region to the Simón Bolívar highway at El Treinta was in disrepair and difficult to travel.

The Simón Bolívar highway itself did not meet the growing transportation requirements of southwestern Colombia. By the late 1940s the port of Buenaventura was handling 43 percent of Colom-

2. The other items purchased for coffee production were *fique* coffee sacks and baskets for picking the berries. The beans were dried in a variety of ways: on coffee sacks spread on the ground, on simple wooden trays laid on the ground, on wooden trays suspended on wooden frames, or on more complicated tiers of trays that could be fanned out or stacked like a set of drawers. The more complicated devices were designed to address the major problem at this stage of the coffee processing: the weather. The coffee harvests, and therefore the drying of the berry, take place in the Dagua mountains during the rainy season. Women and children have the responsibility for bringing in or covering the drying beans when one of the frequent rainstorms approaches. The larger the volume of coffee, the more difficult this task becomes, and the greater the need for moving beyond sacks and trays on the ground (which have to be brought in rather than simply covered). Here, too, however, the solutions generally involved the use of materials available on-farm, with the exception of some of the more complicated tier devices, which had been constructed with the assistance of the Coffee Federation in the 1930s and 1940s.

3. Prior to the FDDZC program, farmers could get advice on coffee cultivation from Coffee Federation extension agents (*técnicos*) in Cali and Dagua.

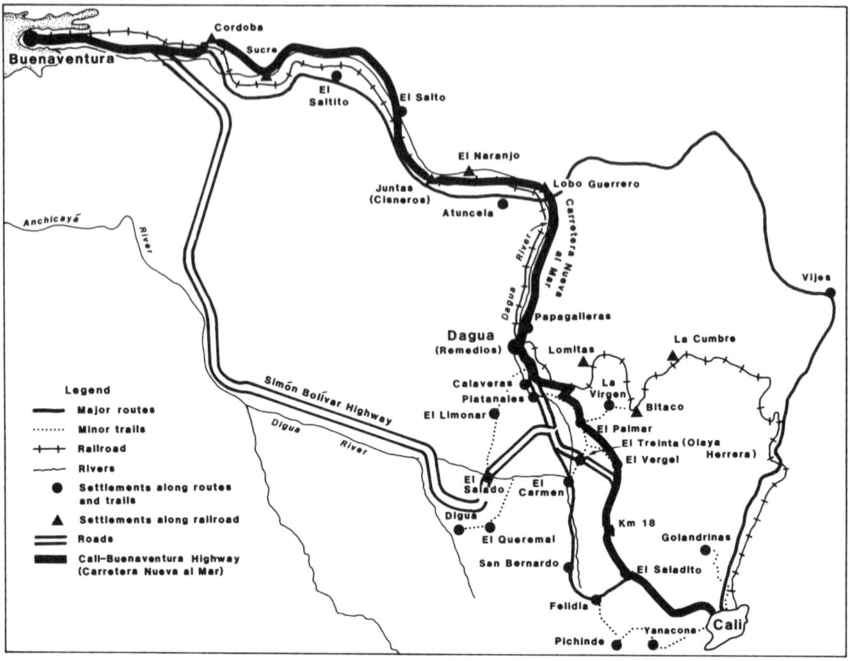


Map 5. The Cali-Buenaventura Highway through El Palmar, 1968.

bia's external trade (McGreevey 1971:258). In the 1950s the tremendous growth of coffee exports and the urban-industrial development of the region, particularly of Medellín and Cali, created new pressures for improved transportation. The Dagua route offered a more direct and rapid connection between Cali and Buenaventura than the roundabout Anchicayá route of the Simón Bolívar highway. In the late 1950s construction was finally begun on a Cali-Buenaventura highway following the old mule-trail route through the center of the Dagua mountains. This route, shown in map 5, was completed in 1968. The trip from Cali to Buenaventura could now be made in three hours. The new Cali-Buenaventura highway (the "Carretera al Mar") and the railroad carried the bulk of the external trade of Colombia's western region. By 1966 almost all Antioquia coffee exports, and 85 percent of all Colombian coffee exports, were shipped out from Buenaventura (Parsons 1968:145).[4]

4. In 1976, 1,393,821 tons of freight moved in and out of Buenaventura, representing 51 percent of the volume of all Colombian imports and exports of that year (CVC 1978:62, table 38).

The bustling new highway passed through the center of El Palmar. For the residents of that community, it was the final step in the slow process of transportation development that had linked them ever more closely to the valley's expanding urban markets. By 1970 Cali was a thriving city of almost one million inhabitants, the third largest city in Colombia (after Bogotá and Medellín). Seventy-seven percent of the population of Valle was urban by 1973 (CVC 1975:6, table 1), and by mid decade Valle produced 19 percent of the country's industrial value-added (CVC 1977a:32).

## Technological Innovations of the 1970s and Farm Income in El Palmar

Economic opportunities for households in the Dagua mountain region expanded in the 1970s as a result of the transportation improvements in the context of regional economic growth, advice and assistance from agricultural extension agents, and a sharp improvement in coffee prices in the mid 1970s (see table 6.15). Agricultural production in El Palmar and neighboring communities began to be "modernized" by the adoption of new crops and the use of nontraditional inputs. An important component of changing agricultural practices was the "technification" of coffee cultivation, supported by the Coffee Federation even in marginal regions during this period of booming coffee prices known as the "coffee bonanza."

### *Household Agricultural Production*

The commercial concentration evident in table 6.1 indicates the extent of the agricultural changes that had taken place in El Palmar by the end of the decade. Four products—coffee, plantains, tomatoes, and *panela*—accounted for 83 percent of the total value of marketed output in 1978–79 for these ninety-two farms. This concentration was evident across all size categories of farm: the four major cash crops (coffee, plantains, tomatoes, and green peppers) accounted for fully 87 percent of the value of marketed output of the smallest units, and 90 percent for the next group of farms. With the exception of the traditional cash crop of *panela*, which remained important for farms of 7.5 or more acres, the farmers of El Palmar were concentrating in 1978–79 on the "new" cash crops of

TABLE 6.1. *Value of Marketed Output by Product, El Palmar, 1978–79*

| | *Acres Farmed in El Palmar* | | | | |
|---|---|---|---|---|---|
| *Product*[a] | *0.1–3.6 (26 farms)* | *3.7–7.4 (22 farms)* | *7.5–29.9 (36 farms)* | *30.0+ (8 farms)* | *Total*[e] *(92 farms)* |
| Coffee | $2,170 (2) | $11,251 (1) | $54,338 (1) | $25,375 (1) | $93,134 |
| Tomatoes | 2,650 (1) | 1,475 (2) | 11,690 (2) | 2,188 (4) | 18,003 |
| *Panela*[b] | 0 | 0 | 4,375 (4) | 8,356 (2) | 12,731 |
| Plantains | 700 (3) | 888 (4) | 7,550 (3) | 3,075 (3) | 12,213 |
| *Arracacha*[c] | 0 | 506 (5) | 2,818 (6) | 1,575 (5) | 4,899 |
| Green peppers | 510 (4) | 900 (3) | 3,185 (5) | 0 | 4,595 |
| Maize | 320 (5) | 120 (7) | 2,703 (7) | 325 (9) | 3,468 |
| Beans | 185 (7) | 113 (8) | 2,633 (8) | 530 (8) | 3,461 |
| Pigs | 275 (6) | 175 (6) | 1,350 (9) | 963 (7) | 2,763 |
| Milk | 0 | 0 | 1,100 (10) | 1,225 (6) | 2,325 |
| Other[d] | 135 | 782 | 3,153 | 2,108 | 6,178 |
| Total[e] | $6,945 | $16,210 | $94,895 | $45,719 | $163,769 |

*Source:* Household interviews by author.
[a] In order by total value, in U.S. dollars. To convert to pesos use exchange rate of 40 pesos to the dollar. Number in parentheses is ranking within each size category.
[b] Unrefined block brown sugar.
[c] A root crop similar in appearance to cassava.
[d] Eggs, pineapples, chickens and other fowl, miscellaneous fruit, cows, cassava.
[e] Figures may not total exactly, due to rounding.

coffee with plantains, tomatoes, and, for the farms with less than 7.5 acres, green peppers.

It will seem strange to the reader that I have classified coffee with plantains as a "new" product. Yet this was the case. Underlying the output figures of table 6.1 were deeper changes taking place in the techniques of production. Coffee cultivation was steadily being transformed by the adoption of *caturra* (the new high-yielding variety of the *arábica* plant), and of other practices being recommended by the Coffee Federation extension agents. By 1978–79, 46 percent of the households with at least .75 acre of land in coffee had planted *caturra*. The adoption rate did not vary systematically by farm size; 44 percent of farms under 3.7 acres that were growing at least .75 acre of coffee had planted some *caturra*, 35 percent of farms between 3.7 and 7.4 acres, 56 percent of farms between 7.5 and 31 acres, and 40 percent of farms over 31 acres. The adoption of this new variety, and the increased cultivation of new crops such as tomatoes and green peppers, involved far more than a simple shift in crops of the kind the farmers of El Palmar had undertaken throughout the century. Rather, the cultivation of these new crops and varieties marked a substantial transition from an essentially closed agricultural system, relying on traditional, nonpurchased inputs (except for some hired labor), to a system heavily dependent on commercial inputs.

A few figures reveal the deeper nature of the transformation that had taken place in El Palmar in one short decade. By 1978–79, 63 percent of the sixty-seven farms with at least .75 acre of land in coffee were using commercial fertilizer on their coffee trees. Every farmer growing tomatoes or green peppers used commercial fertilizer, improved seed, and chemical pesticides for those crops, which are highly susceptible under tropical conditions to pests and diseases. In addition, many farmers were using purchased fertilizer on some of their traditional crops, particularly maize with beans.[5] It should be kept in mind that in the 1960s no farmer in El Palmar used improved seed or pesticides, and only four farmers were using commercial fertilizer for coffee cultivation before 1968.

The differences between the techniques of cultivation for the traditional crops and varieties and those used for the new crops and varieties in El Palmar are presented in detail in Appendix C. In addition to a greater use of complementary purchased inputs, the new crops and varieties required a greater labor input per unit of land cultivated (although not necessarily per dollar of output) than the traditional crops and varieties. As a result of these requirements, the farmers of El Palmar rapidly increased their cash outlays per acre of cultivated land. In order to finance these expenditures, they turned increasingly to the use of bank credit, as table 6.2 reveals.[6] By 1978–79 almost one-third of the households had

5. Most commonly this was chicken manure (*gallinaza*) produced in the valley as an important by-product of the poultry industry.

6. Loans were available to the farmers of El Palmar in 1978–79 from a variety of sources on different terms. The Banco Cafetero in Cali, run by the National Coffee Federation, provided principally medium-term (two- to three-year) loans for the planting of coffee, at 15–18 percent annual interest. The Caja Agraria in

TABLE 6.2. *Credit Use by Households in El Palmar*

| | Acres Owned | | | | |
|---|---|---|---|---|---|
| | *0–3.6 (41 families)* | *3.7–7.4 (21 families)* | *7.5–22.4 (24 families)* | *22.5+ (19 families)* | *Total (105 families)* |
| Never used credit | 20% | 10% | 4% | 21% | 14% |
| Have used credit from: | | | | | |
| Private banks | 2% | 10% | 38% | 32% | 17% |
| Caja Agraria | 7% | 24% | 75% | 42% | 32% |
| Coffee cooperative | 5% | 24% | 54% | 26% | 24% |
| Store credit | 68% | 81% | 54% | 26% | 60% |
| Relatives or neighbors | 49% | 52% | 54% | 47% | 50% |
| Other sources | 17% | 5% | 17% | 16% | 14% |

*Source:* Household interviews by author.
*Note:* Acreage owned includes land outside El Palmar. Totals for each farm size category exceed 100 percent, since households often borrowed from more than one source.

borrowed from the Caja Agraria (the Agrarian Reform Agency Bank) and almost one-quarter from the Coffee Cooperative. Even more significantly, 17 percent had obtained loans from private banks. Not surprisingly, these figures were highest for larger farms. Nevertheless, the use of bank credit had extended to the smaller farms, and even to some of the households with less than 3.7 acres.

---

Cali, an arm of the Agrarian Reform Agency (INCORA), provided similar loans, plus shorter-term loans for the planting of annual cash crops such as tomatoes or string beans, also at 15–18 percent annual interest. The Coffee Cooperative in Dagua, run by the National Coffee Federation, provided loans for home improvement at 13 percent annual interest and small coffee-planting loans at 8 percent annual interest to members only (most of the farmers in El Palmar were or had been members). The local Aqueduct Committee (Junta del Aqueducto) made small short-term loans at 2.5 percent monthly. Some private individuals in Dagua and in the area of El Palmar (including one of the El Palmar shopkeepers) made short-term loans at 5 percent monthly interest. In addition to these sources of cash loans, there were more traditional sources of loans and credit. One of the coffee purchasers in Dagua made advances against the coffee harvest, deducting the "interest" from the price paid for the harvest; some intermediaries advanced cash against other market crops (tomatoes, green peppers, string beans, etc.); the shops in El Palmar generally advanced credit for purchases (*fías*); and neighbors and relatives loaned cash, often without interest.

Indebtedness in 1978–79 was widespread: 45 percent of the sixty-four households with at least 3.7 acres of land were in debt when interviewed. Current debt was quite high in many cases: 52 percent of indebted households owed between $625 and $1,875 (25,000 to 74,999 pesos) and another 17 percent owed $1,875 or more. As a point of reference, note that a fully employed adult male agricultural laborer (*jornalero*) could earn approximately $750 (30,000 pesos) annually in El Palmar in 1978–79.[7] These debt figures do not include store credit, which, as table 6.2 indicates, was quite extensive in the community.

An even clearer indication of the extent of the commercial and technological transformation of El Palmar agriculture in the 1970s was obtained by creating a composite index that aggregates the components of this modernization process: the use of nontraditional inputs, bank credit, the mechanization of various operations and the purchase of certain agricultural equipment, the adoption of new activities such as beekeeping, and so on. Such a composite measure was calculated for the sixty-four households with at least 3.7 acres of land. A detailed specification of the indicators used and the weights assigned to each is presented in Appendix D.

Possible scores on this index ranged from a low of 0 to a high of 230. The results are presented in table 6.3, where the categories can be interpreted approximately as follows: 0 = "traditional" (no mechanization, no use of purchased inputs, no credit, etc.); 5–20 = "improved" traditional coffee cultivation;[8] 25–50 = use of nontraditional inputs for coffee cultivation, or cultivation of nontraditional crops on a small scale but without bank credit; 55–100 = some mechanization, heavy use of nontraditional inputs, moderate to high use of bank credit; 105 or more = fully mechanized modern coffee operation, or moderately mechanized coffee operation plus cultivation of nontraditional crops, heavy use of nontraditional inputs and of bank credit.

From table 6.3 it can be seen that only six out of sixty-four

7. Based on the following figures: twelve weeks at harvest pay (six days per week at 150 pesos per day) and thirty-eight weeks at nonharvest pay (five days per week at 100 pesos per day), for a total of 29,800 pesos per year. Workers employed year-round by the same employer were also, according to Colombian labor legislation, entitled to an annual bonus.

8. Households might also fall into this category if they operated a store, reflecting another type of economic transformation taking place in El Palmar.

TABLE 6.3. *Technological Transformation in El Palmar, 1978–79: Index Scores of 64 Households Owning 3.7 or More Acres*

| *Index Score*[a] | *Number of Households* | *Percentage of Households* |
|---|---|---|
| 0 | 6 | 9.4 |
| 5–20 | 12 | 18.8 |
| 25–50 | 21 | 32.8 |
| 55–100 | 20 | 31.3 |
| 105–230[b] | 5 | 7.8 |
| Total | 64 | 100.0 |

*Source:* Household interviews by author.
[a]For explanation of index see text and Appendix D.
[b]Highest score obtained was 135.

households, or 9 percent, were "traditional" (scored 0). Another twelve, or 19 percent, scored between 5 and 20, indicating "improved traditional" cultivation—that is, slight modification of traditional farming practices. The remaining households, forty-six out of sixty-four, or 72 percent, were moderately to heavily involved in commercial and technological transformation.

The data on agricultural operations collected in the 1978–79 interviews suggest that the adoption of the new farming practices had increased net farm income. Higher net farm income was correlated with the household's cash expenditures for farm operations, which in turn was a good indicator of the household's use of nontraditional crops and technology in the year of the survey.[9] More direct evi-

9. Net farm income as used throughout this chapter was calculated as market sales plus the value of farm output consumed by the household minus production expenses for the agricultural year. Depending on when the household was interviewed, the data cover either the two 1978 growing seasons or the second 1978 season and the first 1979 season. Depreciation was not included in my calculation of production expenses, as I did not have the necessary data on value and life of agricultural capital. The relationship between net farm income and cash outlays was determined in a regression of net farm income on cash outlays, holding constant the amount of land cultivated and the value of the family-farm labor input, for the sixty-four households owning 3.7 or more acres. The coefficient obtained on the cash variable was positive and highly significant. The conclusions presented here as to the interpretation of this statistical result depend, however, on

dence is available from crop studies carried out with several farmers in 1979. Each kept a detailed record for one growing season of all operations for a particular crop. Table 6.4 presents the labor input, variable costs of production, and gross and net income calculated from these records for two cases each of tomatoes, green peppers, and intercropped maize and beans (one traditional case and one using improved inputs). In addition, this table reports data calculated on a retrospective basis for one case each of traditional and technified coffee cultivation.

The data from the crop studies show that markedly higher net income per acre and per day of labor input were obtained from tomatoes and technified coffee cultivation than from the other crops. The green peppers in both cases became diseased, and yields were considerably reduced. Even so, the net income per acre in one of these cases was higher than for traditional maize/bean or coffee cultivation. The data in this table, however, overstate the potential gains from modernization and technification in an important way. Costs of equipment, such as pesticide applicators for tomato and green pepper cultivation and drying and processing equipment for coffee, were not included in my calculations. The type and price of this equipment varied widely and information on the appropriate depreciation period was not available.

The equipment costs involved, especially for technified coffee cultivation, could be considerable. The data presented for technified coffee in table 6.4 were obtained from the farm of Arcesio Benavides and his wife Diomelina Álvarez. This couple incurred a debt of 100,000 pesos ($2,500) to build a new coffee-processing unit (*beneficiadero*) and drying roof capable of handling their increased coffee output. Nevertheless, their income gains from technification were still considerable, and their material standard of living in 1978 was vastly improved over their difficult economic situation in the 1960s. I have presented a more detailed discussion of this family's experience in Appendix E, case 1, as an example of the positive income effects of technification for small coffee cultivators. I suspect that, while including the cost of equipment in table 6.4

---

certain assumptions about the reasons for observed differences in agricultural practices among these households. For a more detailed presentation of the regression procedure and interpretation, see Reinhardt 1987b: 230–33.

TABLE 6.4. *Crop Study Cost and Revenue Data, Individual Farms, 1978–79*

| | *Maize with beans*[a] | | *Coffee*[a] | | *Tomato*[b] | | *Green Pepper*[b] | |
|---|---|---|---|---|---|---|---|---|
| | *Traditional* | *Improved* | *Traditional* | *Technified* | *Case A* | *Case B* | *Case A* | *Case B* |
| 1. Gross revenue | $214.50 | $627.50 | $270.00 | $1,925.00 | $1,335.00 | $1,650.25 | $83.00 | $117.50 |
| 2. Nonlabor variable costs: | | | | | | | | |
| Seed | $11.75 | $36.50 | $1.00[c] | $59.75[d] | $5.00 | $3.75 | $2.50 | $2.50 |
| Fertilizer & pesticides | $5.75 | $12.50[e] | $5.00 | $220.00 | $209.90 | $127.50 | $73.13 | $48.13 |
| Transport | —[f] | —[f] | $0.75 | $5.80 | —[f] | —[f] | —[f] | —[f] |
| Total | $17.50 | $49.00 | $6.75 | $285.55 | $214.90 | $131.25 | $75.63 | $50.63 |
| 3. Net Revenue a: (1) − (2) | $197.00 | $578.50 | $263.25 | $1,639.45 | $1,120.10 | $1,519.00 | $7.37 | $66.87 |
| 4. Labor input, man-days: | | | | | | | | |
| a. Hired labor | 26.00 | 0 | 24.00 | 86.00 | 0 | 6.50 | 1.50 | 0 |
| b. Family labor | 16.25 | 65.00 | 26.00 | 90.00[g] | 124.50 | 110.00 | 30.50 | — |
| Total | 42.25 | 65.00 | 50.00 | 176.00 | 124.50 | 116.50 | 32.00 | — |
| 5. Net return per man-day of labor: (3) ÷ (4) | | | | | | | | |
| Dollars | 4.66 | 8.90 | 5.27 | 9.32 | 9.00 | 13.04 | 0.23 | — |
| Pesos | 187 | 356 | 211 | 373 | 360 | 522 | 9 | — |

| | | | | | | | | |
|---|---|---|---|---|---|---|---|---|
| 6. Cost of hired labor | $65.00 | 0 | $90.00 | $322.50 | 0 | $13.00 | $3.75 | 0 |
| 7. Cost subtotal b: (2) + (6) | $82.50 | $49.00 | $96.75 | $608.05 | $214.90 | $144.25 | $79.38 | $50.63 |
| 8. Net revenue b: (1) − (7) | $132.00 | $578.50 | $173.25 | $1,316.95 | $1,120.10 | $1,506.00 | $3.62 | $66.87 |
| 9. Net return per man-day of family labor: (8) ÷ (4b) | | | | | | | | |
| Dollars | 8.12 | 8.90 | 6.66 | 14.63 | 9.00 | 13.69 | 0.12 | — |
| Pesos | 325 | 356 | 267 | 585 | 360 | 548 | 5 | — |
| 10. Return per *plaza*, net of cash variable costs:[h] (8) ÷ land in crop | $132.00 | $578.50 | $173.25 | $1,316.95 | $4,480.40 | $6,024.00 | $14.48 | $267.48 |

*Source:* Reinhardt 1987b.
[a] One *plaza* (0.64 hectares) in crop.
[b] One-quarter *plaza* (0.16 hectares) in crop.
[c] Value of seed divided over fifteen-year productive life of tree.
[d] Cost of *caturra* seedlings and planting, divided over ten-year productive life of tree.
[e] Land had been in tomatoes and had been heavily fertilized.
[f] Transport costs were deducted from price received, as well as cost of shipping crates for tomatoes and green peppers. Cost of *fique* sacks for coffee, maize, and beans were not included since reused, and depreciation was difficult to determine.
[g] Includes family labor input for seedling preparation and planting averaged over ten-year productive life of tree.
[h] Opportunity cost of family labor, cost of credit, and depreciation value of equipment, buildings, and work animals not netted out.

calculations would have reduced the reported income gains from modernization and technification, they would still have been considerable.

We can also see from table 6.4 that the use of new inputs ("improved" commercial seed and fertilizer) for maize/bean cultivation doubled net income per day of labor and quadrupled net income per acre, compared to traditional practices. This example overstates the net income gain from improved maize/bean cultivation since the land in question had been heavily fertilized under tomato cultivation the previous season and the amount of fertilizer applied to the maize/bean crop was unusually low for improved cultivation. Even had the usual amount of fertilizer been applied, however, there would still have been a significant income gain. This example does show that the income gain from tomato cultivation may have been greater than the direct effect measured in the single-crop data if tomato cultivation was rotated with another annual crop that could benefit from residual fertilizer in the soil.

The data on farm operations suggest that innovating farmers in El Palmar obtained gains in net farm income. In addition, the generally favorable market conditions of the 1970s meant increased income from traditional as well as innovative crops and practices. Farmers expanded the overall level of agricultural output in response to favorable markets and new technology (Reinhardt 1987b: 225–26).

## *Absentee-Owned Farms in El Palmar*

While many family farmers began to adopt new crops and technology, the largest (120-acre) absentee-owned farm in the community continued to be operated under a traditional system of cattle grazing and sharecropping.[10] The valley resident who had purchased this farm in 1954 had let it deteriorate, selling off parcels totaling 135 acres, since he was too busy with his mines to supervise it (and had had "problems with workers and residents of the area," including some land invasions). He left the day-to-day operations in the hands of his *agregados* (a resident manager and his family) and sharecroppers.

10. Information in this section is based on interviews in 1978–79 with nine absentee owners or their farm managers. More information on absentee-owned farms is provided in Reinhardt 1981:328–32.

TABLE 6.5. *Comparison of Traditional and Modern Absentee Farms, El Palmar, 1978–79*

| | Traditional Farm | Modern Coffee Farms[a] | | |
|---|---|---|---|---|
| | | A | B | C |
| Farm size (*plazas*) | 80 | 4.7 | 19 | 40 |
| Cash outlay in year of study | $2,000 | $2,800 | $11,244 | $38,750 |
| Cash outlay per *plaza* cultivated[b] | $50 | $700 | $592 | $969 |
| Coffee yield (*arrobas*/*plaza*) | 20 | 90 | 100 | 80 |
| Gross farm income in year of study | $8,250 | $4,725 | $19,375 | $11,250 |
| Net farm income in year of study | $6,250 | $1,925 | $8,131 | –$27,500 |
| Net income per *plaza* owned | $78 | $410 | $428 | –$688 |
| Net income per *plaza* cultivated | $156 | $481 | $428 | –$688 |
| Current indebtedness | 0 | $1,000 | $12,500 | 0 |
| Gross return per peso of cash outlay | 4.13 | 1.69 | 1.72 | 0.29 |

*Source:* Farm interviews by author.

*Note:* All farms owned by individuals with no kinship ties to the community of El Palmar. To convert dollar figures to pesos use exchange rate of 40 pesos to the dollar. One *plaza* equals 6,400 square meters (0.64 hectares, or approximately 1.5 acres). One *arroba* equals 12.5 kilograms or 25 Colombian pounds.

[a] Farm A had 3 *plazas* of coffee in production and 1 newly planted; Farm B, 9.5 in production and 9.5 recently planted; Farm C, 37 being planted to new coffee and only 3 in production.

[b] Including pasture land.

Although this traditional absentee owner eschewed the new crops and coffee varieties, he did begin to enter into arrangements with his sharecroppers in the 1970s to use modern inputs such as improved seed and fertilizer for their food crops. He provided half these expenses and the land, while they provided the other half and all labor. He did not improve his coffee operation, however, because he feared a labor shortage. This resulted in the low coffee yields shown in table 6.5 for this "traditional" farm, comparable to the yields of some traditional family farmers. Although the dairy, sharecropping, and coffee operations were run with low cash outlays, they generated reasonable returns per acre and high returns per peso of outlays, as table 6.5 demonstrates. The *en companía* arrangements were also lucrative for some of his sharecroppers, as cases 5 and 12 of Appendix E illustrate.

Conversely, other absentee investors purchased land in El Palmar in the 1970s precisely because they hoped to profit from the "coffee bonanza" with technified coffee operations.[11] Table 6.5 presents cost and revenue data for three such farms. The smallest, a modern coffee farm, was managed by an El Palmar resident and doubled as a weekend home for the owner's family, while the other two were highly technified coffee plantations that employed several *mayordomos* (overseers) and large numbers of permanent and seasonal laborers housed in dormitories.[12] These farms were equipped with motor-driven depulpers and hot-air dryers for the coffee beans.

The data for farm C show the extremely high start-up costs for a technified plantation: with only three *plazas* (one *plaza* equals 1.64 acres) of coffee in production and thirty-seven *plazas* newly planted, this farm incurred considerable losses in 1978–79. Farm B had half its land in production and half in newly planted *caturra*, while farm A had three *plazas* in production and one newly planted. These two farms, with nearly identical costs and returns per *plaza* of land and per peso spent, were generating much higher net income per *plaza* than the traditional absentee farm. However, their return per peso spent was considerably less, and the operations required a much greater time commitment on the part of the owners.

Another category of absentee owners were sons and daughters who had migrated from El Palmar and subsequently inherited land in the community. In the past, small inherited plots had generally been left uncultivated or were farmed in a partial and traditional manner by the absentee owners themselves (on weekends, once a month, etc.), by relatives, or, rarely, by an *agregado* family. In the 1970s some of these owners, too, began to renovate their coffee stands and technify their operations.

11. Most of these buyers were merchants or professionals from Cali and nearby Valle cities. The incursion of urban capital into coffee cultivation in the 1970s has been noted in many coffee regions (Fajardo 1983:147–48).

12. The 19-*plaza* farm was on land that had once been part of Agapito Cabrera's holdings.

## Distributional Impact of the Commercial and Technological Changes in El Palmar

Who gained from the economic changes of the 1970s in El Palmar? The classic Marxist analysis predicts that commercial expansion and agricultural innovation will lead to the breakup of the peasant economy. Those households with the most land will be able to take advantage of the new opportunities and prosper, while the majority of rural households will eventually lose out in the competition for land and other resources and become impoverished. They will gradually be forced to sell what land they possess to the richer peasants or, at best, will retain tiny plots of land with which they supplement wage income. Chayanovian analysis does not predict this "social differentiation" of the peasantry. Most Chayanovians do believe, however, that peasant producers will be impoverished by the growing "vertical concentration" of agriculture. Another area of contention in the Colombian case, as we have seen, is the impact of the "technification" of coffee production. A number of analysts believe that this has finally enabled the large grower to be more efficient than the family farmer, and these writers are predicting the elimination of small-scale coffee cultivation in Colombia.

How did the developments of the 1970s affect the households of El Palmar? Did social differentiation take place? Did family farmers find it increasingly difficult to compete in coffee markets as national coffee production became technified? Did vertical concentration drain investible surplus from the community and leave the households as impoverished as they had been in the 1960s?

I have already suggested part of the answer to these questions in my discussion of the positive income effects of agricultural innovation: an investible surplus was clearly accruing to these farmers. As to competition from technified estate production, family farmers in El Palmar did not appear to be adversely affected. They were expanding and improving their coffee operations alongside of the highly technified operations of some absentee owners. The technified family farmer of table 6.4 earned $979 per *plaza* in coffee net of variable costs and the imputed value of family labor, calculated at the maximum value of 150 pesos per day. This compares very well with the performance of absentee coffee farms A and B in table 6.5, which netted $642 and $855 per *plaza* of producing coffee. While the large coffee estates and the Coffee Federation may well have continued to reap the lion's share of Colombian coffee income, family farmers in this region also captured important income gains.[13] Even the traditional peasant coffee farmer of table 6.4 earned a positive net return of $75.75 per *plaza* in coffee after imputing a total wage of $97.50 for family labor. This is, of course, an artificial distinction, since both the imputed wage and the "net" return accrue as cash income to the farm household. It does demonstrate, however, that even a traditional coffee operation could be a "rational" venture in the 1970s, when rationality is calculated in terms of the opportunity cost of the forgone wage.

The increased level of agricultural production in the 1970s in turn generated increased demand for farm labor and sustained relatively high agricultural wages.[14] Wages in El Palmar, while below the level of wages in Colombia's principal coffee regions, were above wages in the southwestern mountains of the country and higher than the national average (Reinhardt 1987b:229). Much of the new income of farmers and agricultural laborers left the community in the form of purchases of farm inputs, services, and consumer goods. But some of it circulated through the community, from farmer to laborer and trader, to shopkeeper and butcher and *cantina* operator and tailor and a variety of other local providers of goods and services. Demand increased for construction workers to build new stores and *cantinas* as well as home and farm improvements. The highway was not only a factor in the expanded farm activity, but also created a whole new set of income-generating activities related to the marketing of products in Cali and in El Palmar itself (Appendix C provides a brief summary of the diverse marketing arrangements for farm products in the late 1970s). While some of these employment and business opportunities were captured by new migrants to the community, many residents also benefited.

Map 6 shows some of the commercial development that had taken place in El Palmar by 1978–79. The roadside market was

13. Extension agents in Nariño and Cauca informed me that this was true in the mountain communities of those departments as well. Palacios (1980:241–47) suggests that peasant coffee producers in marginal coffee areas may generally have benefited from the coffee bonanza of the 1970s.

14. The largest absentee-owned coffee farm (farm C in table 6.5) alone would require over 100 seasonal harvest workers once in full production. The owner was planning to recruit these workers from the nearby communities.

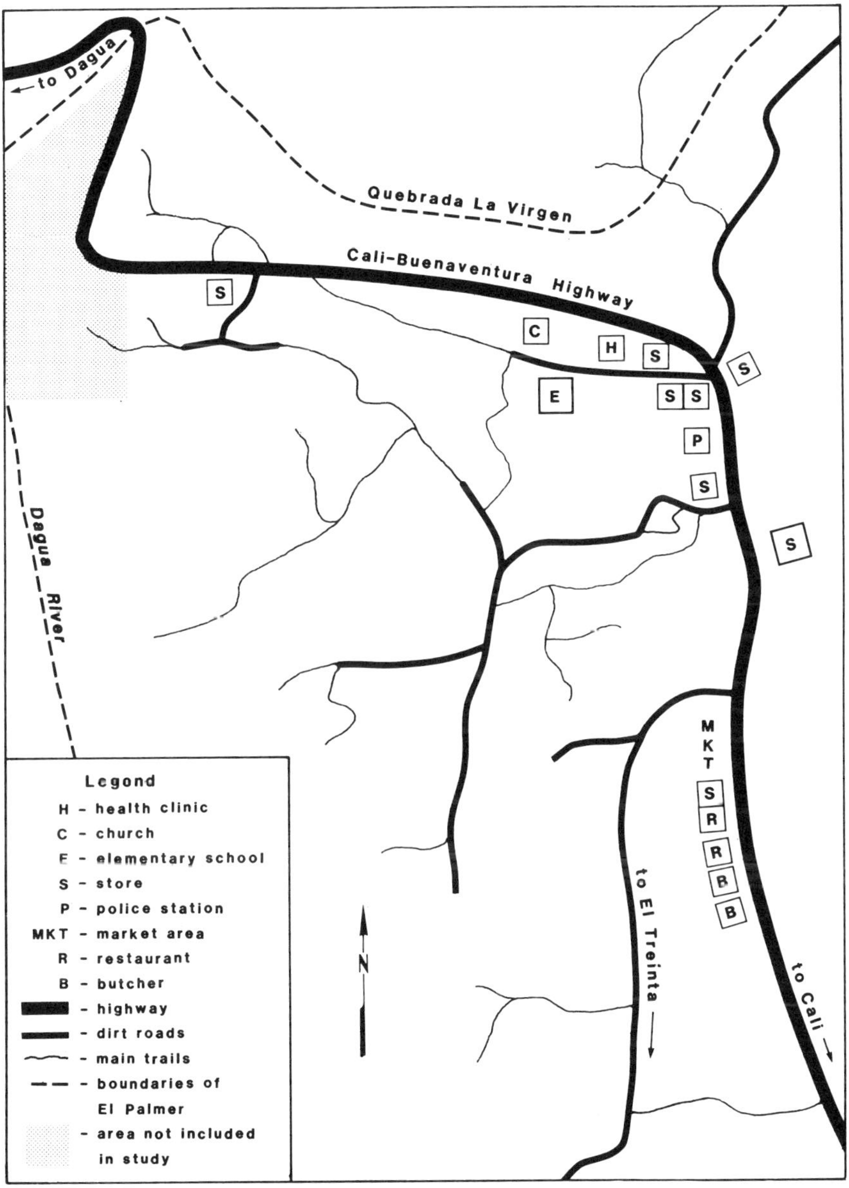


Map 6. The Village of El Palmar, 1979.

thriving. There were seven stores and *cantinas* in operation, with another under construction. Near the market were two restaurants and two butcher shops. There were also a police station (*inspección de policía*), a recently renovated five-grade elementary school, a new health clinic, and a newly painted church. Other striking signs of improved economic conditions visible throughout the community were home improvements and construction, improvements to community trails and roads, the acquisition of electricity and electric appliances, and the construction of varied types of coffee-drying and processing facilities. Forty-six percent of the resident households interviewed had invested in farm improvements between 1968 and 1978, 38 percent had invested in home improvements, and 27 percent had purchased sewing machines, kitchen appliances, televisions, furniture, or other consumer durables.

There were certainly economic differences among the households of the community, including differences in land ownership, as table 6.6 and figure 6.1 show. Twelve percent of the resident households interviewed owned no land in 1978–79, and another 27 percent owned less than 3.7 acres in El Palmar.

There were also significant differences in the ability of the El Palmar households to meet their basic needs, as indicated by a comparison of each household's income in 1978–79 to the income level necessary to provide its members with the lowest-cost nutritious diet.[15] This is a good summary measure of the material standard of living of each household, since it takes account of food prices and the household's demographic composition as well as its nominal income level. The results, presented in table 6.7, show that 21 percent of the households earned less than they needed to prevent malnutrition, while another 21 percent earned only slightly more than they needed for food requirements. In addition, of course, these households needed to devote income to other basic necessities, so that their actual expenditures on food were sometimes seriously below subsistence levels.[16] In 1978–79, then, over

15. This measure was described in chapter 4, note 26, and is explained in more detail in Reinhardt 1981: Appendix B.

16. Families restricting foodstuff expenditures in order to free up resources for other necessities generally increased the relative weight in the diet of low-cost, high-calorie items such as *panela* and rice, or increased the role of that basic staple of the local diet, plantains.

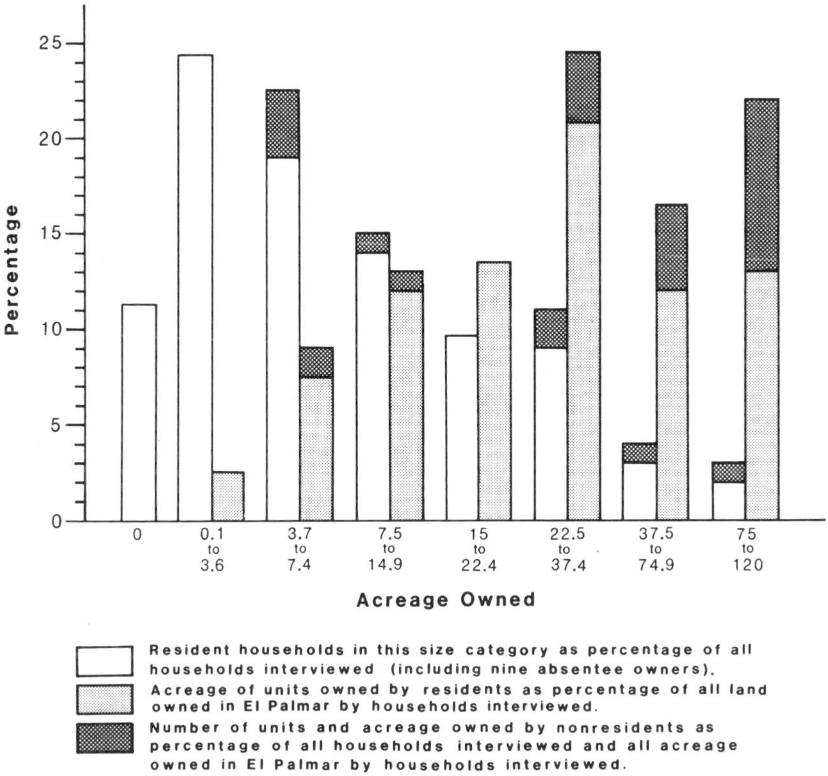


Fig. 6.1. Percentage Distribution of Households by Amount of Land Owned in El Palmar, 1978–79.

40 percent of the households in El Palmar earned an inadequate or barely adequate income.[17]

Table 6.8 gives us a better sense of which households were least able to meet their basic needs in 1978–79. They were households with male heads between the ages of thirty-five and sixty-four who owned fewer than 3.7 acres, households with 3.7 to 7.4 acres

17. In interpreting these figures, it should be kept in mind that they are applicable only to the year of the survey. In an agricultural community, the income of any particular household can fluctuate considerably from year to year. It should also be stressed that few families in El Palmar had any savings to draw on during years of low income. They would sometimes borrow from relatives, neighbors, or moneylenders, often buy from the markets and stores on credit, and generally reduce food consumption to nutritionally inadequate levels during those years.

TABLE 6.6. *Land Ownership in El Palmar, 1978–79, 105 Resident Households*

| | Acres Owned in El Palmar | | | | | |
|---|---|---|---|---|---|---|
| | *0* | *0.1–3.6* | *3.7–7.4* | *7.5–22.4* | *22.5+* | *Total* |
| Households (%) | 12.4 | 26.7 | 20.9 | 26.7 | 13.3 | 100 |
| Holdings (%) | 0 | 30.4 | 23.9 | 30.4 | 15.2 | 100 |
| Land owned (%) | 0 | 3.2 | 9.0 | 32.8 | 55.0 | 100 |

*Source:* Household interviews by author.

TABLE 6.7. *Extent to Which Households Met Basic Subsistence Needs in El Palmar, 1978–79*

| *Net Income ÷ Basic Subsistence Needs*[a] | *Number of Households* | *Percentage of Households* |
|---|---|---|
| Under 0.75 | 9 | 8.6 |
| 0.75–0.99 | 13 | 12.4 |
| 1.00–1.24 | 22 | 20.9 |
| 1.25–1.49 | 18 | 17.1 |
| 1.50–1.99 | 15 | 14.3 |
| 2.00–2.99 | 15 | 14.3 |
| 3.00–4.99 | 8 | 7.6 |
| 5.00+ | 5 | 4.8 |
| Total | 105 | 100.0 |

*Source:* Household interviews by author.

[a]For calculation of basic subsistence needs, see text. Net income is cash farm income plus value of farm products consumed by the household, minus farm costs, plus net income from nonfarm sources. The mean value of this ratio for the 105 households was 2.01.

TABLE 6.8. *Ratio of Net Household Income to Basic Subsistence Needs for 88 Male-Headed Households, El Palmar, 1978–79*

| *Age of Head of Household* | *Acres Owned* 0–3.6 | 3.7–7.4 | 7.5–22.4 | 22.5–37.4 | 37.5+ | *Total*[a] |
|---|---|---|---|---|---|---|
| 25–34 | 1.7<br>(12) | 1.3<br>(2) | 1.1<br>(1) | 3.7<br>(2) | —<br>(0) | 1.9<br>(17) |
| 35–39 | 1.4<br>(5) | 0.9<br>(1) | 3.2<br>(3) | 6.0<br>(1) | —<br>(0) | 2.3<br>(10) |
| 40–49 | 1.1<br>(9) | 1.5<br>(6) | 3.0<br>(3) | 1.5<br>(1) | —<br>(0) | 1.5<br>(19) |
| 50–64 | 1.2<br>(5) | 1.2<br>(4) | 1.8<br>(11) | 7.0<br>(3) | 4.4<br>(3) | 2.5<br>(26) |
| 65+ | 2.0<br>(4) | 1.1<br>(4) | 1.7<br>(3) | 2.2<br>(3) | 2.0<br>(2) | 1.8<br>(16) |
| Total[b] | 1.5<br>(35) | 1.3<br>(17) | 2.1<br>(21) | 4.2<br>(10) | 3.4<br>(5) | 2.0<br>(88) |

*Source:* Household interviews by author.

*Note:* Figures shown are mean values for each group. Figure in parentheses is number of cases in cell. Acreage owned includes land outside El Palmar.

[a]Analysis of variance significance at 50 percent level.

[b]Analysis of variance significance at 99.9 percent level.

whatever the age of the head of household, and households headed by men in their forties, except for three of these men who had farms between 7.5 and 22.4 acres. Female-headed households, not shown in table 6.8, were a mixed group whose economic status depended more on household composition than on the age of the head or land ownership.

These figures show that the average economic conditions of the age cohorts followed a Chayanovian pattern, with the exception of men in their forties. However, the low significance level of this column in the table indicates that there was considerable variation within each age cohort, with some households doing well and others very poorly. The figures also show that ability to meet basic needs in 1978–79 varied significantly by land ownership, with land-poor households on the whole doing less well than households with larger holdings.

Nevertheless, it would be premature to conclude that land-poor households had been impoverished by the economic changes of the 1970s. Similarly, it would be hasty to conclude that those households with the larger holdings in 1978–79 had been the better-off households in the late 1960s. We need to look more carefully at the previous conditions of these households and at how they had participated in the economic changes of the decade. Overall, those changes had impeded rather than encouraged social differentiation in the community by providing new avenues of upward economic mobility for land-poor households. These avenues included farming, agricultural employment, and trade.

### *Changing Status of Land-Poor Households in El Palmar*

Households with fewer than 7.5 acres of land relied heavily on off-farm sources of income in 1978–79, as we can see from table 6.9.[18] This was particularly true for households with holdings under 3.7 acres. Only 10 percent of these households earned over 35 percent of their incomes from their holdings, and none earned over 75 percent. This group used farm production to supplement other income sources. Households with 3.7 to 7.4 acres relied more on farm income. Twenty-three percent of these households earned over 75 percent of their income from their farm operations, while another 64 percent earned from one-third to three-fourths of their income from farming. For this group, nonfarm income supplemented their generally insufficient farm income.

Table 6.10 provides more detailed information on how the sixty-three households with under 7.5 acres were earning a living. This table shows the major source of income for each household, where "major" was defined as providing over 50 percent of household income. The income source was classified as "mixed" if no single activity provided over 50 percent of household income.

Approximately half of the thirty-nine households in table 6.10 with under 3.7 acres relied on agricultural wage labor as their major income source, including in this category households relying on sharecropping arrangements. Eight other households were farm

18. Income obtained from sharecropping (*en companía*) arrangements was considered off-farm income.

TABLE 6.9. *Proportion of Household Income from Farm by Size of Holding, El Palmar, 1978–79*

| *Acres Owned in El Palmar* | *Number of Households* | *Percentage of Households Obtaining from Farm* 0–4% of Total Income | 5–35% of Total Income | 36–75% of Total Income | 76–100% of Total Income |
|---|---|---|---|---|---|
| 0 | 13 | 100 | 0 | 0 | 0 |
| 0.1–3.6 | 28 | 34 | 55 | 10 | 0 |
| 3.7–7.4 | 22 | 5 | 9 | 64 | 23 |
| 7.5–22.4 | 28 | 0 | 7 | 50 | 43 |
| 22.5+ | 14 | 0 | 21 | 14 | 64 |
| Total | 105 | 22 | 22 | 31 | 25 |

*Source:* Household interviews by author.
*Note:* Total net household income includes value of farm produce retained for household consumption.

TABLE 6.10 *Resident Households' Major Income Source by Size of Holding (under 7.5 Acres), El Palmar, 1978–79*

| *Major Source of Income* | *Acres Owned in El Palmar* 0–3.6[a] | 3.7–7.4 | Total |
|---|---|---|---|
| Store | 4 | 0 | 4 |
| Farm manager/overseer | 8 | 0 | 8 |
| Agricultural labor[b] | 19 | 5 | 24 |
| Nonagricultural labor | 5 | 1 | 6 |
| Mixed | 3 | 6 | 9 |
| Farm | 0 | 10 | 10 |
| Total | 39 | 22 | 61 |

*Source:* Household interviews by author.
[a]Not included: two female-headed households, supported by men living elsewhere.
[b]Includes sharecropping.

managers. Nevertheless, nonagricultural sources of income were significant. Four households had taken advantage of the highway to set up small general stores, which provided a major portion of their income. Two others derived some income, although less than a major portion, from small stores or from trade in agricultural products (selling by the roadside, selling in El Treinta, etc.). Nonagricultural labor was the principal income source for five households and provided some income for four others. The main type of nonagricultural labor for the men was construction work, although one man worked for the railroad. Women in several of these households earned income from roadside marketing activities, dressmaking, piecework, factory employment, or paid domestic work.[19]

The majority of the twenty-two households with 3.7 to 7.4 acres relied on farm production or agricultural wage labor as their major income source. None were primarily shopkeepers or farm managers. Members of this group did, however, supplement their primary income source with income from dry-goods stores, agricultural trade, construction work, and other commercial and service activities.

This examination of income sources shows that the land-poor households had not been excluded from the expanding economic activity of the decade. Their participation had extended beyond agricultural wage labor to a variety of other activities, some of which were quite lucrative. Case 3 in Appendix E describes the favorable situation encountered by the household of a farm manager, while case 4 presents the story of a landless laborer who became a successful *cantina* owner. Households dependent on wage labor also found that their situation had improved. Agricultural employment was far more regular than it had been in the depressed conditions of the 1960s. No male over fourteen in the interviewed households

19. In addition to various nonhousehold activities, women in all these land-poor families were responsible for the household chores and for the tending of a small food plot if the family had any land. An important task was to prepare the noon meal for men of the household working in the area and to bring it to them at their work site (day laborers receive higher wages if meals are provided by their families rather than by the employer). In these activities the women were helped by the children of the household. Both women and children helped to tend the store if there was one. By the age of fifteen, children in these households were expected to make a full contribution to family income: males working as agricultural laborers, females sometimes being sent to Cali as domestic servants.

reported difficulty obtaining as much wage employment as he desired, despite the rapidly expanding working-age population of the community (the figures in table 5.1 showed that the number of offspring of the resident households who reached the age of fifteen was far greater than the number of working-age offspring who migrated between 1968 and 1978; both seasonal and permanent migration into the community also increased). Nevertheless, land-poor households relying primarily on shopkeeping or employment as farm managers were better able to meet their subsistence needs in 1978–79 than those relying primarily on wage labor. All twelve households in the former category earned income in 1978–79 sufficient for their basic needs, while only 60 percent of the thirty households relying primarily on wage labor were able to do so.[20]

In addition to accruing the positive effects of the expanding income-earning opportunities of the decade, land-poor households benefited directly from the favorable market conditions for agricultural products. As table 6.6 indicates, households with under 7.5 acres owned 12.2 percent of the land in the study. Some of these households also rented or sharecropped land, so that they operated 16 percent of the land in the study. Although some agricultural output was consumed directly by the households,[21] the major portion of their production was sold, as we can see from table 6.11. For the smallest farms, the value of marketed output accounted for 65 percent of the value of total farm production, while for the next category of farms that figure rose to 76 percent.

20. It is sometimes argued, as we saw in chapter 1, that household subsistence production enables wage workers from smallholding families to accept a lower wage than would otherwise be the case, creating a subsidized or "semiproletarian" labor force. Whether the going wage in El Palmar was below subsistence depended on household composition. As pointed out earlier, a fully employed adult male could earn approximately 30,000 pesos in 1978–79. This was considerably greater than the approximately 9,500 pesos that he needed to obtain a nutritionally adequate diet (migrant workers paid 200 pesos weekly, or the equivalent of 10,400 pesos a year, for their cooked meals). However, it was barely adequate to cover the annual food needs of a couple with two young children (approximately 27,500 pesos), and clearly inadequate to this family's overall subsistence needs. While farm income could provide a necessary supplement to the father's wage income, the economic situation of households relying primarily on wage labor would improve only as offspring began to reach working age and could contribute to the household's cash income.

21. Appendix C discusses production for household consumption by El Palmar farm families.

TABLE 6.11. *Marketed Output versus On-Farm Consumption, El Palmar, 1978–79*

| | Acres Farmed in El Palmar | | | | |
|---|---|---|---|---|---|
| | *0.1–3.6 (26 farms)* | *3.7–7.4 (22 farms)* | *7.5–29.9 (36 farms)* | *30.0+ (8 farms)* | *Total (92 farms)* |
| Marketed output[a] | $ 6,945 | $16,210 | $ 94,895 | $45,719 | $163,769 |
| Output consumed on-farm[a] | $ 3,663 | $ 5,025 | $ 13,038 | $ 4,913 | $ 26,639 |
| Total output[a] | $10,608 | $21,235 | $107,933 | $50,632 | $190,408 |
| Marketed output as % of total | 65 | 76 | 88 | 90 | 86 |
| Ratio, marketed to consumed output | 1.90 | 3.23 | 7.28 | 9.31 | 6.15 |

*Source:* Adapted from Reinhardt 1983: Table 1.
[a]Figures are total for farm-size category, in U.S. dollars. To convert to pesos use exchange rate of 40 pesos to the dollar.

It is important to stress that small farmers as well as large participated in the income gains from agricultural innovation. We saw in table 6.1 that tomatoes and green peppers had become important cash crops on the small farms. The technical characteristics of these crops made them well suited to small-scale, family-labor-based cultivation. The susceptibility of tomatoes and green peppers to pests and disease under tropical conditions meant that continual supervision and care were necessary. Because of the requisite supervision, the smaller farms with smaller acreage devoted to these crops had the highest yields, as can be seen from table 6.12. As a result of these diseconomies of scale, plot sizes for these crops tended to be quite small, less than one-half acre on average, and they had been adopted by many of the land-poor households.

Another important factor in the adoption of these crops by small farmers was the high income per acre that they could potentially generate, as the crop studies indicated. We can see from table 6.13 that farmers with less than 3.7 acres of land devoted almost 40 percent of their agricultural land to traditional foodstuffs, the bulk of which were consumed by the farm household. Tomatoes and green

TABLE 6.12. *Average Yields and Acreage of Tomatoes and Green Peppers by Size of Farm, El Palmar, 1978–79*

| | *Acres Farmed in El Palmar*[a] | | | |
|---|---|---|---|---|
| | *0.1–3.6* | *3.7–7.4* | *7.5–29.9* | *30.0+* |
| Tomatoes: | | | | |
| Average crates per acre | 855 | 712.5 | 315 | 400.5 |
| Average acres sown | 0.45 | 0.30 | 1.95 | 1.20 |
| Green peppers: | | | | |
| Average crates per acre | 213 | 525 | 198 | 0 |
| Average acres sown | 0.90 | 0.45 | 1.05 | 0.30 |

*Source:* Household interviews by author.
*Note:* Averages are for farms growing these crops.
[a] Including rented or sharecropped land.

peppers enabled these land-scarce farmers to increase their cash income considerably with only a minimal diversion of land away from foodstuffs. In 1978–79 tomatoes and green peppers accounted for 30 percent of the total value of agricultural output on farms under 3.7 acres, but used only 10 percent of their agricultural land. As competition for land increased in the 1970s, the ability to farm their small plots more intensively was an important improvement for land-poor households.

While these new crops were suited to land scarcity, two basic requirements for their cultivation were labor and financing. Households with fewer than 3.7 acres were heavily dependent on income from wage labor. Nevertheless, they were able to combine day labor during the high-wage period of the coffee harvest with tomato or green pepper cultivation during the slack period between coffee harvests, since the seasonal pattern of activities for annual crop cultivation in El Palmar was complementary to the labor requirements of coffee cultivation (Reinhardt 1981:385–89).[22] Farmers in the 3.7 to 7.4 acres category, however, were generally stretching family labor to the utmost in their attempts to earn a livelihood from their holdings. As table 6.13 shows, these farmers chose to

22. As larger farmers expanded their technified coffee operations, however, they found that the off-peak labor demands of technified coffee cultivation severely restricted their ability to cultivate other crops.

TABLE 6.13. *Land Use in 1978–79 by Size of Operation, 92 Farms in El Palmar*

| | *Acres Farmed in El Palmar*[a] | | | | |
|---|---|---|---|---|---|
| | *0.1–3.6 (26 farms)* | *3.7–7.4 (22 farms)* | *7.5–30.9 (36 farms)* | *31.0+ (8 farms)* | *Total (92 farms)* |
| Total acres in agricultural use | 32.75 | 81.50 | 423.50 | 175.75 | 713.50 |
| Percentage in: | | | | | |
| Pasture | 4.6% | — | 9.0% | 35.0% | 14.2% |
| Coffee[b] | 45.9% | 76.1% | 55.8% | 37.1% | 53.1% |
| Sugar cane | 0.5% | 1.3% | 6.4% | 11.5% | 6.8% |
| Traditional foodstuffs[c] | 38.1% | 18.9% | 21.3% | 14.8% | 20.1% |
| Tomatoes | 4.6% | 1.5% | 4.2% | 1.3% | 3.2% |
| Green peppers | 5.5% | 1.8% | 2.1% | 0.2% | 1.8% |
| Other fruits & vegetables | 0.9% | 0.4% | 1.1% | 0.1% | 0.8% |

*Source:* Reinhardt 1987b.
[a]Includes rented or sharecropped land.
[b]With plantains (not reported separately). Also, annual crops intersown with coffee are not reported separately.
[c]Various combinations of maize, beans, cassava, and *arracacha*.

concentrate on coffee cultivation, leaving only a small proportion of land and labor for tomatoes and green peppers. Even so, that 3.3 percent of their agricultural land generated 11.2 percent of the total value of their output.

Financing was not a difficulty for land-poor households wishing to cultivate tomatoes or green peppers. Cultivators could get advances against their crops from traders. Young men in the community with no land of their own also entered into *en companía* arrangements with relatives, neighbors, and absentee owners. Through these arrangements they obtained access to land and varying proportions of the necessary finances, and they obtained the rest from traders or from their own savings. Because of the generally prosperous market conditions, landless farmers were able to use these arrangements advantageously, as Appendix E illustrates, sometimes cultivating traditional foodstuffs for the market as well as tomatoes and green peppers.

The evidence I have presented so far has been primarily static, related to the situation in 1978–79. We can get a better sense of the impact of the economic changes of the 1970s by looking at the changing conditions of individual households over the decade. One important element of those changing conditions was the movement of households from one land ownership category to another. The classic Marxist analysis reviewed in chapter 1 suggests that concentration of land ownership will be the result of economic development, with many farmers losing access to land while the richest increase their farm size. Chayanovian analysis, as we saw in chapter 4, predicts that young households will gradually expand their landholdings as they progress through the family life cycle, while older households diminish holding size as offspring leave the family unit, resulting in a fairly stable distribution of land ownership over time.

Changes in land ownership in El Palmar during the 1970s conformed more to the Chayanovian than to the Marxist pattern. Many young households were able to buy land, primarily from the older large farms, whose owners were attempting to reduce farm size commensurate with shrinking family size, or from large landowners who wanted to get out of farming in El Palmar altogether. Young men were obtaining funds to purchase land from their work as farm managers, traders, and sharecroppers, as well as from loans. Two of Lubino Cabrera's sons had been able, through income from sharecropping and with a commercial loan, to purchase sizeable holdings of their own before they reached their mid thirties. Their story is presented in Appendix E, as case 5. Another young man had started by purchasing a tiny plot of land with his savings from agricultural labor, on which he grew tomatoes. He had then turned to agricultural trade, using the market buses to transport produce from El Palmar to Cali and to bring back the supplies (fertilizer, pesticides) his growers needed. He had been able to purchase a large farm in El Palmar with his earnings and a commercial loan. His story is presented as case 2 in Appendix E.

These were not isolated cases of upward mobility. As table 6.8 indicates, seven households with male heads under forty years of age had managed to acquire over 7.5 acres during the 1970s, with only one obtaining that land through inheritance. We saw in chapter 4 that the economic downturn of the 1960s had impeded upward mobility and magnified the significance of family background for the cohort of men forming households in that period. The favorable economic conditions of the 1970s reversed those trends, with young landless households finding a variety of independent means to earn a livelihood in the community and to obtain the resources with which to buy land.

Not all households in El Palmar had found the conditions of the 1970s to be so propitious. In general, the households doing most poorly in 1978–79 were those whom the changes of the 1970s had bypassed or not yet reached: younger families just starting out (especially those with many young children), men in their forties who had entered the 1970s with large young families and little or no land, and those older households who lacked both land and family labor. Men in their forties had found it particularly hard to overcome their difficult start in the previous decade. Case 7 in Appendix E describes the conditions of Carlos Cabrera and his family in the 1970s as an example of the problems faced by these households. Twenty-nine percent of the men in their forties owned between 3.7 and 7.4 acres in 1978–79, which table 6.8 shows to be the land ownership category with the lowest average standard of living. As we have seen, these households tended to concentrate on income from farming, supplementing that income with off-farm earnings. With little time to devote to nonfarm activities, these households were least able to take advantage of the new income-earning opportunities of the 1970s. They were also the most cautious about experimenting with agricultural innovation, given the importance of every bit of cultivable land to household income.

One question, however, is whether the changes of the 1970s actually placed some households at a disadvantage, particularly those with little land. Food and land prices had risen. Did this result in declining real wages and reduced access to land for households dependent on wage income? The evidence suggests that the impact of these changes was mixed.

Only eleven of the sixty-three households with less than 7.5 acres of land in 1978–79 felt that their economic situation had deteriorated over the previous decade. Of these, eight were older households in which most or all of the children had left; three of these eight were headed by elderly widows. These households eked out a living from their holdings (most in the 3.7–7.4 acre

TABLE 6.14. *Economic and Demographic Characteristics of El Palmar Households, 1978–79*

| *Economic Situation 1968–78 (no. of households)* | *Average YNET/SUB*[a] | *Average Age of Head of Household* | *Percentage of Heads of Household 60 or Older* | *Percentage of Female-Headed Households* | *Average No. of Acres Owned, 1968* | *Average No. of Acres Owned, 1978* | *Percentage Owning <2.5 Acres, 1978* |
|---|---|---|---|---|---|---|---|
| Improved (48) | 2.6 | 45 | 12.5 | 4.2 | 5.7 | 9.0 | 35.4 |
| No change (30) | 1.4 | 49 | 26.7 | 23.3 | 7.5 | 9.7 | 46.6 |
| Worsened (27) | 1.1 | 60 | 48.1 | 25.9 | 16.0 | 13.0 | 22.2 |

*Source:* Household interviews by author.
[a] Ratio of net household income to basic subsistence needs; see text.

range), sometimes supplemented with agricultural wage work. Their farms were poorly tended, in traditional foodstuffs and old coffee. The rising cost of hired labor, given the low productivity of their farms, had led them to reduce or eliminate the use of *jornaleros*. The other three households who felt their conditions had deteriorated were headed by younger men who had large families with no children old enough to make a financial contribution. The difficult conditions faced by these households is illustrated by case 6 in Appendix E. Although members of these two groups felt that their conditions had worsened, the majority of land-poor households felt that their situation had remained unchanged or had improved.

Evidence from the community as a whole suggests that it was predominantly the older households who felt that their conditions had worsened over the 1970s, regardless of farm size. As table 6.14 shows, the twenty-seven households who felt that their situation had deteriorated were markedly older than those households who felt their position to be unchanged or improved. That the deterioration was unrelated to access to land is indicated by the fact that these twenty-seven households had larger holdings on average than the rest of the households. In general, those households placed at a disadvantage by the changes of the 1970s were households with rising dependency ratios (nonworking to working members) for whom rising food costs and/or rising costs of hired labor outweighed the benefits of rising crop prices or improved employment opportunities.[23] It is difficult to say what the situation of these households would have been had the economic turnaround of the 1970s not occurred. The condition of the cohort of men in their forties compared to that of the two younger cohorts suggests that, on the whole, economic conditions would have been worse for the younger households and perhaps also for some of the elderly households.

## The Family-Farm Labor Process in the Changing Environment of the 1970s

While the commercial and technological developments of the 1970s were favorable to the economic status of El Palmar households,

23. Female-headed households were also disproportionately represented among households who felt that their situation had not improved or had de-

they also had important implications for household labor relations and the labor process on the family farm. The household dynamics observed by this author in El Palmar suggest that the economic cohesion of the patriarchal units was being undermined. Although emigration had always been a drain on the family labor force, the developments of the 1970s drew even the resident family members in diverse directions. Individual family members were increasingly involved in independent local economic activities. This was a particularly notable change for teenaged girls and women, whose extrahousehold activities in the 1970s included fast-food stalls, agricultural trade, shopkeeping, dressmaking, raising small fowl, and both agricultural and nonagricultural employment in El Palmar and the neighboring communities.[24] There was also increased off-farm economic activity among young men.

---

teriorated. Many of these women had been recently widowed or were attempting on their own to maintain a family of young children.

24. There is considerable debate about the effect of economic development on the economic status of women. While much of this literature has focused on urban employment and particularly female factory work, there has been some debate as well about the effects of agricultural modernization on rural women. Very little of this literature has dealt with the types of development taking place in El Palmar; the focus has tended to be on the direct and indirect effects of commercial agricultural development, or on the effects of industrial competition that destroys rural handicraft work traditionally performed by women. Much of this literature has emphasized the negative effects of these changes on women's economic status (their reduced income-earning potential due to competition from larger, more efficient producers; the increased degree of rural male migration straining family ties and leaving women as sole providers for themselves and their children; sex discrimination that reserves the better-paying jobs for male employees); or the arduous, dangerous and low-income character of the new employment opportunities available to them. The classic work dealing with the question of the overall effect of economic development on women's economic status is Boserup 1970; later studies in this vein include Haggag Youssef 1974; Tanker and Bramsen 1976; Beneria 1982; and ECLA 1975 and 1983, the ECLA studies dealing specifically with Latin America. Examples of studies examining the effect of commercial competition on rural women's economic activities in Latin America are Deere 1977 and Deere and León de Leal 1982; these studies generally conclude that rural women's economic status is reduced by these types of development processes. One study that examines changes in Andean women's economic status as a result of increased commercial contact with urban areas (a process akin to that taking place in El Palmar) finds evidence of increased participation in trading activities, although concluding that women's overall economic position had deteriorated relative to that of males (Bourque and Warren 1981). In El Palmar, too, women benefited from their participation in the economic changes of the 1970s, although the impact on their relative status is unclear. In some cases the patriarch channeled part of the increased farm income into home improvements that directly

The improvement in local economic conditions changed the relative bargaining positions of father and son. Whereas there was greater likelihood that sons would stay in the community to take advantage of the new economic opportunities, it was now necessary for the patriarch to offer greater incentives to maintain the son's direct participation in family-farm activities. These incentives included *en companía* arrangements for tomatoes and green peppers in which the patriarch provided both land and the necessary cash; early inheritances; and wage payments to sons for their on-farm work. All of these arrangements meant that the sons would be available for full- or part-time work on the patriarch's farm at the same time as they gained some economic independence. These arrangements often implied greater innovation and adoption of new crop/technology packages, since the sons tended to be less wedded to traditional practices and also needed to generate greater income from smaller plots than their fathers. These arrangements also, however, often implied fragmentation of the family labor process and a loss of control by the patriarch over the family labor force and the overall organization of family-farm production.

These changes in the availability of both male and female family labor occurred precisely when demand for labor was increasing and agricultural tasks were becoming more time-constrained. Although the value of family labor was enhanced by the changes of the 1970s, the greater economic independence of family members began to undermine the economic advantages of the patriarchal labor process. The family farmers turned increasingly to the use of hired labor, and efficiency was reduced by shortages of labor at key periods in the production cycle.[25] While patriarchal control of the family-

benefited the wife (wood or cement floors instead of hard-packed dirt, better-sealed walls, new kitchens, appliances that made some household chores easier, electricity, sewing machines, etc.). More often, the benefits to the wife derived from her direct participation in the new income-earning opportunities of the period.

25. For example, the young coffee plants had to be transplanted to the field within a fairly narrow time frame. If this period were allowed to pass without transplanting, the roots of the plants would pass through the plastic bags and lodge in the soil where they stood. Thereafter the plants were unlikely to survive transplanting. In 1978 Aristides Benavides lost 2,000 coffee plants because he was not able to obtain labor for planting during the necessary two-week interval, which fell during the coffee harvest. Timing was also important for the traditional

TABLE 6.15. *Internal Coffee Prices for Café Manizales and Consumer Price Index, Colombia, 1970–78*

| | Internal Coffee Price in Colombian Pesos[a] | Nominal Coffee Price Index (1975 = 100) | Consumer Price Index, Workers (1975 = 100) | | Real Coffee Price Index (1975 = 100) |
|---|---|---|---|---|---|
| | | | In Cali[b] | Nationwide[c] | |
| 1970 | 130.4 | 47.8 | — | — | — |
| 1971 | 128.4 | 47.0 | — | — | — |
| 1972 | 149.7 | 54.8 | 55.4 | — | 99.0 |
| 1973 | 192.9 | 70.7 | 66.0 | — | 107.1 |
| 1974 | 220.7 | 80.8 | 82.6 | — | 97.9 |
| 1975 | 273.0 | 100.0 | 100.0 | 100.0 | 100.0 |
| 1976 | 650.0 | 238.1 | 123.1 | 117.4 | 202.8 |
| 1977 | 730.0 | 267.4 | — | 168.5 | 158.7 |
| 1978 | 730.0 | 267.4 | — | 190.6 | 140.3 |

*Sources:* Coffee prices from DANE 1978; consumer price index, Cali, from DAP 1978:170; consumer price index, nationwide, from FEDECAFÉ 1978.
[a] Per *arroba* (one *arroba* = 12.5 kilograms).
[b] Average for each year.
[c] June of each year.

farm labor process had not been completely eroded, its weakening and the increased use of hired labor did reduce the traditional importance to family-farm competitiveness of its distinctive labor process.

## Weathering the International Downturn of the Early 1980s

It is important to note that the positive income effects of technological and commercial development in El Palmar were due to

crops. However, strategies had been devised to minimize the labor constraint for these crops. For example, maize and beans were left to dry in the field if the necessary labor was not available for harvesting when these crops were ripe. This tied up the land for an additional period and reduced the farmer's income; nevertheless, it yielded a positive net return on the farmer's investment. The root crops common to the area had no harvest constraint: they would keep for a lengthy period in the soil, barring pest infestations. Here again, the land was tied up until the harvest, but the crop did not perish. With coffee, tomatoes, and green peppers, the crop would be lost if harvesting did not take place when the fruit was ripe.

the favorable market environment of the 1970s. High coffee prices and the related regional economic growth were the umbrella under which the local peasant economy could flourish. Table 6.15 shows that internal coffee prices in the 1970s rose more rapidly than the regional and national consumer price indices.

By the late 1970s, however, that umbrella was beginning to fold. International coffee prices declined with the recovery of the Brazilian plantations, and internal coffee prices, after stagnating for several years, were lowered in mid 1979 (CIEP 1979). As a result, major lending institutions began to cut back on loans to the coffee areas (Martin 1979). At the same time, all indications were that input costs would continue to rise. Pesticides and fertilizer would become more necessary in El Palmar as soil quality continued to deteriorate and pest resistance increased. Labor costs could also be expected to increase as new coffee trees in El Palmar reached full production levels and more labor was required. When I left El Palmar in 1979, it appeared that farmers might be caught in a cost/price squeeze precisely when loan funds dried up.

Surprisingly, when I returned to El Palmar in December of 1983, I found that coffee farmers had done well in the intervening years.[26] Although international coffee prices had continued to decline, as table 6.16 shows, internal prices had been raised from 730 pesos per *arroba* (12.5 kg sack) in 1978 to 1,410 pesos in December 1983. This was facilitated, as we can see from table 6.16, by the steady devaluation of the peso. While external coffee prices fell by

26. Information on conditions in El Palmar is based on a follow-up study of the community from December 1983 through January 1984. During this visit, information was obtained on farm input prices, output prices, land prices, and food costs. It was determined that 9 of the original 105 households had left the community, and four male heads of household and two female heads of household had died. Of the four nondescendant absentee owners, one had died, one had sold out, and a third had sold his small farm and bought a larger one in the community. In order to obtain a general idea of the economic conditions in El Palmar in the years since 1978–79, a sample of the 105 households was reinterviewed. The sample was selected by dividing the households into groups according to major source of income and, for farmers, type of farm operation, in 1978–79. The categories were: landless agricultural laborer, *agregado/mayordomo*, shopkeeper or other roadside *negocio*, middleman, small farm mixed-income, diversified farm (low and high debt), coffee farm (low and high debt). Representative households were selected for the follow-up interviews within each category. In addition, general information on households not reinterviewed was obtained from informants and brief visits to some households.

TABLE 6.16. *Colombian Coffee Prices, 1977–83*

| | *Internal Price*[a] | | *External Price*[b] | | *Peso Exchange Rate* | | *External Price Converted to Pesos*[c] | |
|---|---|---|---|---|---|---|---|---|
| | *Pesos* | *Index (1977 = 100)* | *U.S. $* | *Index (1977 = 100)* | *Pesos per U.S. $* | *Index (1977 = 100)* | *Pesos* | *Index (1977 = 100)* |
| 1977 (Dec.) | 7,300 | 100.0 | 2.03 | 100.0 | 37.96 | 100.0 | 77.06 | 100.0 |
| 1978 (Dec.) | 7,300 | 100.0 | 1.31 | 64.5 | 41.00 | 108.0 | 53.71 | 69.7 |
| 1979 (Dec.) | 8,300 | 113.7 | 1.82 | 89.7 | 44.00 | 115.9 | 80.08 | 103.9 |
| 1980 (Dec.) | 9,200 | 126.0 | 1.26 | 62.1 | 50.92 | 134.1 | 64.16 | 83.3 |
| 1981 (Dec.) | 10,330 | 141.5 | 1.43 | 70.4 | 59.07 | 155.6 | 84.47 | 109.6 |
| 1982 (July) | 11,050 | 151.4 | 1.31 | 64.5 | 64.69 | 170.4 | 84.74 | 110.0 |
| 1982 (Oct.) | 12,100 | 165.8 | 1.41 | 69.0 | 66.42[d] | 175.0 | 93.65 | 121.5 |
| 1983 (July) | 12,800 | 175.3 | 1.27 | 62.6 | 80.00 | 210.7 | 101.60 | 131.8 |
| 1983 (Nov.) | 14,150 | 193.8 | 1.45 | 71.4 | 88.77[e] | 233.9 | 128.72 | 167.0 |

*Sources: Revista Banco de la República* 1983:95 (official exchange rates, December 1978 to July 1983) and 125 (internal and external coffee prices, 1977 to April–July 1983); *La República*, 2 December 1983 (Bogotá), p. 1; *El Siglo*, 30 November 1983, p. 11; IMF 1985:154–55 (official exchange rates, December 1977 to December 1983).

[a] Pesos per 125 kg sack of premium-grade parchment coffee (official Colombian Coffee Federation price).

[b] Dollars per 453.6 g (U.S. lb.) of "other milds," New York Commodities Exchange.

[c] External price per U.S. lb. multiplied by current peso exchange rate.

[d] September exchange rate.

[e] December exchange rate.

almost 30 percent between 1977 and 1983, the value of the peso fell by over 50 percent in that period. This devaluation alone would have increased the Coffee Federation's peso income from external coffee sales from 77 to 128 pesos per pound, despite the external price fall, as the last two columns of table 6.16 indicate. This reduced the cost to the Federation of raising internal coffee prices.

Devaluation to raise export producer prices is frequently proposed as a solution to agricultural export stagnation in many Third World countries. As frequently occurs with currency devaluations, however, Colombia experienced considerable inflation between 1978–79 and 1983–84. This inflation can be harmful to both farm and nonfarm households, as prices of purchased foods and farm inputs rise. Table 6.17 presents data on price changes for farm inputs, agricultural outputs, and principal purchased foodstuffs in El Palmar between 1978–79 and 1983–84. From this table we can see the rapid inflation that occurred over the five-year period. The national consumer price index rose from a base of 100 in 1978 to 274 in 1982–83 and 318 in 1983–84 (Zambrano 1986). As the real income of large coffee growers began to fall, they put increasing pressure on the Coffee Federation to raise internal coffee prices. The Federation responded with price increases that were consistently greater than its income gains from devaluation, as shown by a comparison of the second and last columns in table 6.16. Nevertheless, these increases did not keep pace with inflation.

Although real coffee prices declined, the overall umbrella for continued prosperity in El Palmar had been kept in place with the devaluation of the peso. Increased coffee yields associated with the new coffee plantations throughout the village partly compensated for the falling real price, and growers' net income position remained favorable. Coffee continued to be profitable in El Palmar, as cases 8, 9, and 10 in Appendix E illustrate,[27] and credit continued to be available. Households reported that lending agencies had allowed them to renew outstanding loans, although with an interest penalty, and there had been no cases of farm loss due to loan default. Far from impoverishing peasant producers, the "vertical concentration" of coffee marketing in Colombia had protected their incomes from the full impact of the international price fall in this period.

27. A severe drought in 1983, however, sharply reduced farm income.

TABLE 6.17. *Price Changes, El Palmar, 1978–79 to 1983–84*

| | *Peso Price, 1978–79* | *Peso Price, 1983–84* | *Percentage Increase* |
|---|---|---|---|
| Producer prices: | | | |
| Coffee | 600–700/*arroba* | 1,200–1,400/*arroba* | 100% |
| Plantains[a] | 50/bunch | 200/bunch | 300% |
| *Panela*[b] | 120–150/*arroba* | 500–600/*arroba* | 300–316% |
| Beans, *guarzo*[c] | 375–425/*arroba* | 700–1,250/*arroba* | 86–194% |
| Maize[d] | 100–125/*arroba* | 500/*arroba* | 300–400% |
| Tomatoes | 100–300/box | 400–1,300/box | 300–333% |
| Green peppers | 80–220/box | 200–800/box | 150–264% |
| Pineapple[e] | 40–50 each | 70–80 each | 60–75% |
| Input prices: | | | |
| Daily wages: | | | |
| With meals | 80–100 | 200 | 100–150% |
| Without meals | 120–150 | 300 | 100–150% |
| Fertilizer: | | | |
| Urea | 500/bag | 1,100/bag | 120% |
| *Producción* | 400/bag | 800/bag | 100% |
| Organic[f] | 50/bag | 250/bag | 400% |
| Land | 35,000/*plaza*[g] | 120,000/*plaza*[g] | 243% |
| Purchased food prices: | | | |
| Rice | 10–12/lb | 25/lb | 108–150% |
| Beans | 16–25/lb | 50/lb | 100–213% |
| Plantains[a] | 50/bunch | 200/bunch | 300% |
| Beef | 40–50/lb | 160/lb | 220–300% |
| Potatoes | 4–6/lb | 12–20/lb | 200–233% |

*Source:* Prices observed by author.
*Note:* One *arroba* equals 12.5 kilograms.
[a] Average-size bunch (one stem).
[b] Unrefined block brown sugar.
[c] *Pintado*, not completely dry.
[d] *Choclo*, still moist; sold on the ear.
[e] Price of sale in Cali.
[f] Processed chicken manure.
[g] Price of one particular plot. Land prices in El Palmar varied sharply in each period based on quality of soil, steepness, and proximity to highway and community roads.

The maintenance of internal coffee prices sustained regional development. Farmers and traders found that demand in the urban markets for such crops as tomatoes and green peppers kept pace with increasing supply. As table 6.17 shows, prices of the principal farm products in El Palmar kept pace with input prices and with the prices of commonly purchased foods. Agricultural prosperity sustained high levels of employment for agricultural laborers and supported continued commercial development in the community. A number of new stores, *cantinas*, butcher shops, and fast-food stalls had opened between 1979 and 1984.

Despite these gains, the food prices and wage data shown in table 6.17 suggest that households dependent entirely on agricultural wage labor experienced a real income decline over the period. However, the positive employment effects of the continued expansion of coffee production and commercial development must be balanced against the negative effects of inflation on wage income. The cases of several households, presented in Appendix E as numbers 11, 12, and 13, show that the experience of the 1979–84 period was generally favorable for land-poor as well as landed households. Cases 14, 15, and 16 illustrate the generally prosperous conditions of shopkeepers and traders in this period.

## Lessons from El Palmar: The Impact of Commercial and Technological Development on Peasant Producers

Two important conclusions emerge from the experience of El Palmar households in the 1970s and early 1980s. The first is that commercial and technological development does not necessarily impoverish peasant households. On the contrary, this development can bring new resources into the peasant community to substitute for the declining land base and complement the growing labor force, as occurred in El Palmar. As a result, the local economy was able to support a larger resident (and nonresident) population.[28] Although young men and women continued to leave El Palmar in the 1970s, the local economy was once again providing opportunities for many young households to establish themselves. The experience of El Palmar thus provides support to the position argued by Echavarría (1976), Corchuelo (1979), Esteva (1978), and others that economic development in Latin America can provide a basis for the persistence, rather than the elimination, of smallholding producers.

28. In addition to the growing number of descendants living in El Palmar, the economic boom of the 1970s has also attracted many "outsiders": transient families who occupy shacks by the highway, seasonal laborers who come for the coffee harvest, and more permanent agricultural workers.

Fundamental to this outcome in El Palmar were the technological and institutional factors that rendered the new income-earning opportunities suitable to small-scale operations, thereby allowing poor households to participate as well as those with more land or financial resources. Barriers to entry were low because of the weak scale economies in the new crop-technology packages, and because of institutional arrangements such as *en companía*, trading on consignment, and crop advances. We cannot conclude, however, that these institutional arrangements would necessarily have the same favorable impact in other communities. In cases of market power, they can be a means of transferring producer surplus to those who control financial and marketing resources. That poor households in El Palmar were able to benefit from these arrangements was due to the relatively egalitarian nature of the community and the competitive environment in which they were operating. Similarly, we cannot assume that technological innovations will continue to be suitable to small-scale, family-labor-based production processes. I shall return to this issue in chapter 7.

The second conclusion is that commercial expansion and technological innovation, even in favorable circumstances such as those of El Palmar, may not be sufficient by themselves to resolve the rural problem. We saw that poverty was still widespread in El Palmar in the late 1970s. This poverty did not reflect a transitional situation that would be alleviated as the benefits of modernization spread, but rather the inherent limitations of modernization given the conditions under which these households attempted to earn a livelihood (Reinhardt 1987b). Poor peasant families, living increasingly in a context of land scarcity and economic uncertainty, seek strategies that will minimize the possibility of failing to achieve subsistence, rather than strategies that will maximize income but are subject to considerable risk.[29] The new crop-technology packages in El Palmar fell into this latter category, capable of considerably increasing income, but with high income variance, as is illustrated with the green pepper cases in table 6.4.

29. This view that peasant "profit maximization" is constrained by risk considerations differs both from the Chayanovian view of drudgery limitations on farm expansion and from the view that peasants are unconstrained profit maximizers. A detailed analysis of the impact of risk on the decisions of the small Colombian coffee grower is presented by Ortiz 1973.

The high income variance and riskiness stemmed from several factors: the considerable price variability for the new perishable crops (during one six-week harvest period tomato prices in Cali varied from as high as 240 pesos to as low as 75 pesos a crate) and the extremely high cash and labor investments they required, as well as their susceptibility to pests and diseases (see Appendix C); the need to take land out of cultivation for eighteen months in order to plant *caturra* (given the credit stipulation that no food crops be interplanted with the coffee seedlings—a particularly difficult decision for farmers with little land) and difficulties encountered by El Palmar farmers due to the ecological unsuitability of several of the specifications for technified coffee cultivation; and the use of bank credit to finance these activities.

Indebtedness was a risky undertaking for households that were generally ill-equipped to manage their debts. Few adults in the community had received more than three years of schooling, and none had any training in the rudiments of debt management, let alone accounting. No accounts were kept, and few households had cash savings. Surplus was sometimes held in the form of coffee beans withheld from sale at the time of harvest, to be sold whenever cash was needed. With no concrete plan for loan repayment, families tended to meet payments haphazardly. Many households in the community had become involved in a rotating system of indebtedness in which they continuously borrowed from new sources to repay old debts. Although they had been able to refinance and no families had defaulted, a number were finding themselves heavily burdened by rising interest payments, as the cases in Appendix E illustrate. Several farmers were considering selling part or all of their land to pay their debts. Others were turning to wage labor in an attempt to earn additional cash income to cover their interest payments.

As a result of the technical and financial difficulties that they had encountered, many farmers had pulled back from technological innovation. Of the eighty-six farmers with over .1 hectare of land, 42 percent had experimented with tomato cultivation in the 1970s. Of these, 8 percent had reduced their cultivation of tomatoes and another 47 percent had stopped growing this crop altogether. For green peppers, 29 percent of El Palmar landowners had tried this crop, and 44 percent of these experimenters had subsequently dropped it. The principal reasons given for dropping these two crops were production problems (insects, disease, drought, illness from chemicals) or that cash outlays were "too high" (as table 6.4 demonstrates, this was relative to the possibly low earnings from these crops rather than relative to potential peak earnings). The experience of these farmers was a negative demonstration effect to others, who said they did not plan to grow these crops. Those who did cultivate these crops tended to do so, as we have seen, on a very small scale.

Coffee technification had similarly encountered a limited receptivity in El Palmar. Despite the considerable efforts of Coffee Federation agents and the demonstrated income gains from technified coffee, over half of the coffee growers interviewed still had no land under *caturra* in 1978–79, and over one-third did not purchase fertilizer for this crop. There were no significant differences by farm-size category in the rate of adoption of *caturra*. Rather, coffee growers in each farm-size category had experienced difficulties with strict adherence to technification recommendations (Reinhardt 1987b), so that farmers in El Palmar were pulling back from full-scale technification in favor of small experiments with modified *caturra* cultivation (such as *caturra* mixed with other coffee varieties, with greater spacing, with more shade trees, and with interplanted food crops) or experiments with the use of commercial fertilizer on traditional coffee varieties.

In general, the farmers of El Palmar were selectively integrating elements of the technological innovations into their existing agricultural systems. In addition to the new crops and varieties and their associated inputs, farmers were experimenting with the use of fertilizer on traditional crops (maize/beans, coffee). With selective adoption, they achieved some income gains with minimal risk; in doing so, they forfeited the greater gains that might have been achieved with a riskier strategy of full-scale technological innovation. It is likely that the extent of adoption would have been even lower if market conditions had not been as favorable as they were in El Palmar in the 1970s.

These conclusions are supported by the results of the Colombian rural development programs initiated in the 1970s by ICA and later the DRI. As we saw in chapter 5, these projects were the principal government response to the dual problems of inelastic domestic food production and social tensions resulting from rural poverty. The objective of the programs was to increase the marketing of basic foodstuffs by raising the productivity of smallholdings. Both the ICA and DRI projects focused on the coordinated provision of technical assistance, inputs, and credit to smallholders in selected communities. This assistance was planned to reach some 92,000 families under Phase 1 of the DRI program. In addition, the DRI projects included expenditures on rural infrastructure such as roads, aqueducts, and electricity, and on training, health, and education, which would benefit a larger number of families. These infrastructural and social components were to complement the technical aspects of the program through their effects in increasing productivity (Tobón C. 1981; Dirección General DRI–PAN 1982b; División de Evaluación PAN–DRI 1982).[30] Overall, the elements of the DRI projects looked very much like the technical, infrastructural, and social changes that took place in a less coordinated fashion in El Palmar.

Evaluations of the Phase 1 DRI projects concluded that the results had varied considerably (Tobón C. 1981; Valencia González 1982; Dirección General DRI–PAN 1982b; Latin American Regional Report 1982). The projects had been generally unsuccessful in two kinds of communities: ones where households owned very small plots and spent most of their time working off-farm, and ones where households had access to land only through yearly rental arrangements. The principal DRI evaluation (Valencia González 1982) argued that producers in these communities (located mainly in the Atlantic Coast districts and parts of Boyacá) had not responded to DRI recommendations because of risk aversion, as indicated by their high degree of product diversification and limited degree of market participation. Although the report emphasizes the risk aversion of the producers rather than the nature of the DRI packages, it is clear that, as occurred in El Palmar, the new practices were indeed riskier than traditional practices. The evaluation notes production problems with the DRI recommendations, ranging from weak corn stalks unable to support climbing beans to serious pest and disease problems. As in El Palmar, the new tech-

30. A good analysis of the overall strategy of the DRI program is provided by Fajardo 1983: 129, 135, 137–41.

niques necessitated considerable cash outlays, while market prices tended to be highly variable.

The characteristics of the producers rather than those of the DRI package were considered to be the problem in these communities because of the apparent success of DRI projects in communities where producers had larger average holding size and/or longer-term sharecropping arrangements, and greater proximity to urban markets. In this second group of communities, adoption rates were higher, default rates lower, and participants had significantly increased their output. The report suggests that producer income increased as a result of the projects.[31] The conclusion drawn from these results was that the packages were successful in areas where risk aversion was lower, and the report recommended that Phase 2 projects be limited to these communities.

However, the degree of success even in these communities should not be overstated. There were cases where inadequate transportation, the lack of storage and processing facilities, and inelastic demand resulted in declining income for farmers who had increased production for urban markets.[32] The overall level of adoption of the technology packages by DRI recipients was low. As in El Palmar, it was more common to find participants selectively integrating elements of these packages into their existing production systems. Even in the "successful" communities, producers attempted to reduce the risk inherent in the new practices by adopting them on a limited scale and emphasizing the least risky components of the technology packages, particularly fertilizer.[33] The results of the DRI projects therefore support the conclusions drawn from the experience of El Palmar: that there can be income

31. This conclusion was based on higher yields and gross income for participants compared to control groups. However, no comparison was made of the initial yield, output, and gross income of the participants and the control groups, nor of their net income at the end of Phase 1.

32. According to *Latin American Regional Report* (1982:2), the "Achilles heel of the [DRI] programmes is marketing, and in some areas improved yields have led to over-production and lower prices because of the distribution bottleneck." While the income of participating households in Antioquia increased with greater production, in Boyacá increased potato production led to net losses for participating potato producers.

33. De Janvry (1981:248) found a similar response by participants in small-farm modernization programs in Mexico and Peru. See also Clawson and Hoy 1979 and Redclift 1983.

gains from modernization packages, that these gains will be greatest where market conditions are most favorable, but that these gains will be limited by the tendency of peasant producers to attempt to reduce the income variance of high-income packages. The comparison of the favorable experience of land-poor households in El Palmar with the negative experience of comparable households in DRI communities also indicates the importance of institutional factors in assuring the greatest range of participation in these income gains.

## The Colombian Peasant Producer in the 1970s: A Summary

Conditions for Colombian peasant producers had changed in the 1970s. Land had become increasingly scarce as the rural population grew and large estates continued to turn to capital-intensive commercial crop production. Economic development weakened patriarchal control over the family labor force at the same time as technological innovation increased both the overall labor requirements of the family farm and the timeliness of farm operations. Unlike in the 1960s, however, difficulties in access to land and control over family labor were countered by more favorable market conditions. The coffee boom generated a new phase of economic development, which benefited family farmers in many regions of the country through expanded markets and improved transportation facilities. The FDDZC program brought new inputs and services to smallholders in the coffee zones, encouraging the "modernization" and "technification" of their operations. In addition, the government sought the "modernization" of peasant production outside the coffee zones through its integrated rural development projects.

Although these factors were not able to resolve the problem of absolute poverty in the Colombian countryside, they did fortify the role of peasant households as agricultural producers. The more favorable conditions of the 1970s enabled many rural households to reestablish the Chayanovian life-cycle pattern of land acquisition. Agricultural production therefore continued to be an important component of an increasingly complex peasant livelihood strategy involving multiple income sources.

*Chapter Seven*

# Whither the Colombian Family Farm?

How are we to interpret the current status of the Colombian peasantry? We have seen that small-scale agriculturalists continue to play an important role as producers of a number of agricultural commodities. At the same time, many of these households are indeed "semiproletarian" in the sense that they combine agricultural commodity production with other economic activities, including wage labor. They are neither pure "peasants," frequently defined as full-time agriculturalists (Wolf 1966; Foster 1967), nor pure "proletarians."

The history of the settlers of the Dagua mountains has shown us, however, that this multiple character of the rural household is not a new phenomenon. Young men have traditionally relied on a variety of income sources in their attempts to establish themselves as family farmers. The majority of those who currently remain in the countryside continue to seek off-farm income not simply as a means of earning a livelihood, but also as a means of establishing or reinforcing their position as agriculturalists. This was evident in the recent histories of many of the young men of El Palmar, and was found generally to be the case in a 1970s study of Colombian agricultural workers in the central Cauca Valley (Bergquist 1986:372–74).

## A Framework for Analysis

The key question then becomes whether or not these men will be able to constitute themselves as either part-time or full-time family farmers over the course of their lives. Their prospects are shaped by an interplay of macro- and micro-level social and economic factors that influence their access to land and markets, control over family labor, and technical competitiveness. This complexity requires us to adopt a broad framework for our analysis. A variety of theoretical and methodological approaches provide important elements for such a framework.

The Chayanovian analysis, for example, focuses our attention on the relevance of the distinctive behavioral characteristics of the family farm to the economic competitiveness of those production units. However, this approach emphasizes only one aspect of that behavior: the willingness of the family farmer to forgo rent, profit, and even the market return to family labor. This analysis stops short of considering the nature of intrahousehold relations and their effect on the family-farm labor process.

The analytical approach of the household economics literature (Folbre 1986), which examines the age- and gender-based household division of labor in different socioeconomic contexts, has offered an important means of enriching the Chayanovian model of the family-labor farm. This analysis draws our attention to the mechanisms of labor organization and control characteristic of the patriarchal household that can increase the efficiency of family labor. As we have seen in the case of El Palmar households, these mechanisms can include the "carrot" of participation in increased farm income as well as the "stick" of various negative economic and social consequences that can result from disobeying patriarchal authority. In addition to an understanding of the organization of the family-farm labor process, a detailed analysis of intrahousehold relations has helped us to understand the demographic behavior of the traditional farm household, with its important consequences for the ability of subsequent generations to establish themselves as agriculturalists.

External factors also influence the intergenerational trajectory of the family-farm economy. The classic Marxist analysis focuses our attention on the impact of market and technological development on agrarian structure. However, this perspective generally dismisses the possibility that small-scale production can be economically competitive with capitalist agriculture, and it overlooks the important differences in the technical characteristics of agricultural as opposed to industrial production. While capitalist agriculture has indeed developed rapidly in Latin America, we cannot assume that this development necessarily signals the destruction of small-scale commodity production.

The persistence of agrarian diversity suggests that we examine more closely the technical character of different organizational forms of agricultural production and the nature of the markets in which they trade. This approach integrates into a broader analytical framework the methodology of traditional agricultural economics, which has tended to eschew a theoretical approach to the analysis of agrarian change trends in favor of empirical measures of relative economic efficiency.[1]

The development of a broader theoretical framework for the analysis of agrarian diversity has been the goal of a number of scholars working within the "mode of production" perspective. This approach has involved attempts to define a variety of "noncapitalist" modes, or ways of organizing economic activity in production units that have different goals and internal relations than those specified as characteristic of "capitalist" units of production.[2] A number of writers have then attempted to understand the ways in which these noncapitalist modes are affected by the development of capitalism, analyzing the factors that have led to the transformation, elimination, or persistence of the noncapitalist modes in this process.[3] Because of the debates that have raged over the theoretical rigor of these uses of the mode-of-production concept, many analysts have dropped this terminology, but continue in the spirit of this method-

1. This includes the work of neoclassical economists and economic anthropologists considering the "efficiency" of peasant producers in terms of their allocation of resources and responsiveness to market price signals, as well as the work of agricultural economists measuring the extent of economies of scale in different agricultural activities.

2. Harrison (1977), for example, reconstructs Chayanov's analysis of the family-labor farm as the specification of a "peasant" mode of production, an approach elaborated on by Warman (1980). Others have rejected the concept of a "peasant" mode in favor of a "simple commodity" mode (Servolin 1972; Bartra 1974, 1975; Ennew, Hirst, and Tribe 1975), an "independent" mode (Kelly 1979), or an "independent family farm" mode (Reinhardt 1981). Bradby (1975) refers to "natural economy."

3. The best examples of this type of analysis are Rey (1973) and Dupre and Rey (1973), who introduce the concept of the "articulation" of modes of production. A recent application of the articulation approach to the analysis of agrarian change in Latin America is the work of Mallon (1983) on Peru.

ological approach.[4] It is no doubt clear to the reader that my analysis of Colombian agricultural structure falls into this category. I have found this to be a useful framework with which to draw together the elements of the various partial analyses into a comprehensive approach to the analysis of agrarian structure.

In examining the changing status of Colombian family farmers, I have emphasized two aspects that have received insufficient attention in the literature on the "peasant question": the issue of limited scale economies and the technical competitiveness of small-scale agricultural production, and the issue of the ability of the peasant household to organize and control the labor force in a manner that preserves its technical competitiveness. It is these factors, as well as broad changes in the conditions of access to land, labor, and product markets, that have determined the possibilities for the intergenerational reconstitution of the Colombian family farm. This chapter considers the potential effect of these factors.

## Population Growth and Access to Land

It is likely that the peasant land base will continue to be limited in Colombia. Although colonization schemes are still being implemented, these projects have had little impact on the overall prob-

4. Hindess and Hirst (1977) and Vergopoulos (in Amin and Vergopoulos 1974) have argued that the specification of a mode must include the political, legal, and ideological institutions and values necessary for the maintenance over time of the constituent units of production. The necessity for this entire fabric of social constructs, referred to as the "social formation," precludes the coexistence of different modes except in a transitional period when a new mode is supplanting a former mode. De Janvry (1981) and others have argued that the peasantry is a class within the "feudal" mode of production, which preceded the development of the capitalist mode. As the institutions of the feudal mode are supplanted by those of the capitalist mode, the peasantry are eliminated as a social category (transformed into members of the capitalist social categories, becoming wage laborers or capitalist producers on a petty scale). While those rejecting the "articulation of modes" approach generally posit the supremacy of capitalist market relations ("relations of exchange"), not all agree that this process is transforming the peasantry into either capitalists or wage laborers. A number of writers have emphasized the continuity of a "form" of agricultural production based on household labor, and have focused their analysis on the ways in which this production unit is integrated into capitalist markets, although reaching widely varied conclusions regarding the resulting class position of the members of these units (Amin and Vergopoulos 1974; Vergopoulos 1975, 1978; Banaji 1977; Friedman 1978, 1980; Schejtman 1980; Goodman and Redclift 1982; Roseberry 1983; Smith 1984a, 1984b; Long 1984).

lem of peasant land scarcity. While pressure for land reform persists, the modernization of estate production has added economic obstacles to the political obstacles already faced by reform efforts.

Rural-urban migration has relieved some of the mounting pressure on the peasant land base, but the rural population has continued to grow rapidly. In earlier chapters we saw that population growth and a resulting expansive drive were essential characteristics of the traditional family-farm economy. Relations between the patriarch and his wife and children gave rise to what might be called an internal "demographic imperative." Can we expect this imperative, and its attendant difficulties, to persist?

Fertility rates have been markedly and persistently higher in rural than in urban areas of the Third World as well as of the advanced industrial nations. This has led scholars to emphasize the fertility-reducing impact of the structural shifts that accompany economic development. As households move into urban settings, as economic activity increasingly takes place in an industrial setting, and as education becomes an important requirement for economic advancement, the "costs" of children are postulated to increase relative to their "benefits," and couples therefore reduce fertility. The literature points in particular to the increased opportunity cost of women's time as they become involved in economic activities outside the home that are incompatible with childrearing tasks. This is contrasted with the greater ease in traditional agricultural economies of combining women's economic activities with reproductive tasks.[5]

The impact on fertility of economic development is not confined, however, to urban areas. Development may also affect the demographic behavior of the rural patriarchal household through its impact on intrahousehold social and economic relations. One issue is

5. Another way of viewing this process is that parents in an urban-industrial setting invest in child "quality" rather than child quantity. See Reinhardt 1987a for a review of the literature on the relationship between economic development and rural fertility. The demand theory of the economic determinants of demographic behavior has been developed by, among others, Becker (1960, 1965), Becker and Lewis (1974), Nerlove (1974), T. W. Schultz (1974), Willis (1974), Easterlin (1975, 1980), and T. P. Schultz (1981). The literature on the negative correlation between the opportunity cost of women's time and fertility in urban areas, and the less clear relationship in rural areas, is reviewed by Abbott (1974), Birdsall (1976), and Standing (1978, 1983).

that of changes in the net economic contribution of children to their parents' household. Much of the literature examining the impact of economic development on the rural household economy points to the erosion of children's economic contribution with the extension of rural education and the increased opportunities for migration of offspring.[6] These factors do not, however, necessarily have a negative effect on household income. As we saw in chapter 4, education and migration can sometimes be a means of increasing children's lifetime contribution to their parents' household, and were selectively used by both Aristides Benavides and Juan Álvarez as part of their overall family-livelihood strategies.

During the 1950s and 1960s, as the schoolhouse was expanded from one room to five separate grades, primary education became increasingly common in El Palmar. In the 1970s the highway facilitated transportation to the secondary school in Dagua, and growing regional prosperity led to the construction of a new secondary school in the nearby mountain community of Bitaco. The increased regularity and length of school attendance did reduce the participation of children in agricultural and household tasks. Nevertheless, the overall economic contribution of children was still significant, both their direct participation during childhood and their later remittances of income earned off-farm. While the number of hours of child labor time may have declined, the changes of the 1970s increased the value of each remaining hour. The data in table 6.4 show the greater net return to labor with the modern crops and technology. Regional economic development also enhanced the earnings potential of off-farm activities.

The patriarch adopting more labor-intensive technology with a greater need for timeliness of operations, facing rising wages and periodic labor scarcity, and seeing greater prospects for the gainful employment of his offspring, would continue to be interested in high fertility.[7] This is particularly the case if increased costs of rear-

6. Writers who have emphasized the importance of children's economic contribution in traditional rural economies include Mueller (1972); Nerlove (1974); Rosenzweig and Evenson (1977); Bulatao (1978); Nag, White, and Peet (1978); Ho (1979); Caldwell (1982). Parsons (1984) argues that increased schooling reduces parental control over household assets since there is no mechanism to enforce the repayment by the offspring of the parents' investment in their "human capital" formation.

7. Several studies have found that increases in the productivity of rural labor were associated with higher fertility (Stys 1957; Rosenzweig 1976).

ing children are borne by other members of the household, as occurred in El Palmar. Mealtime protocol protected the patriarch from any effects of increased family size on nutrition: the male head of household was served first and always received the choicest portion of protein (meat, eggs, beans), while the adult and teenaged females ate last and bore the brunt of any protein deficiency (generally their diets consisted of a much greater proportion of rice).

Despite these factors, fertility in El Palmar, which had begun to decline through family limitation practices in the difficult period of the 1960s, continued to be limited by many couples in the relative prosperity of the 1970s (Reinhardt 1987a). This was a response to several characteristics of the social and economic environment of this period. First, while it was not new for women to bear disproportionately the costs of rearing children, the socioeconomic changes of the 1970s increased those costs. Women were primarily responsible, through their independent cash-income activities, for the educational expenses of their offspring (school fees, books, clothing, transportation). In addition, some of the new economic activities affected the opportunity cost of women's time in the same manner as has been noted for urban settings. These factors clearly changed the cost-benefit calculus of many women in favor of reduced fertility. These changing preferences were reinforced by contact with small-family norms through radio, television, and the actions of family planning personnel.

At the same time, the economic changes of the period helped to create a social and economic context in which women had a greater ability to translate their fertility preferences into practice. Income-earning opportunities increased women's economic autonomy and improved their bargaining position with their spouses.[8] Several women in the community had been able to finance their sterilization operations or the use of other birth control measures such as IUDs. Overall, where women had some independent income sources, the children appeared to be fewer, better clothed and fed, and better educated.[9]

8. A UN Development Program study of rural women's economic participation (UNDP 1980) found similar trends in Haiti as a result of women's growing trading activities.

9. Their independent income was frequently used by women in El Palmar as a shield between the children and the labor demands placed on them by the patriarch. This was particularly the case when it came to choices between the chil-

Conceptions of women's roles and rights, which always lag behind the reality of women's position, were also beginning to change. Although patriarchal authority over wife and home was still generally accepted, and frequently enforced through the physical abuse of women, there were cases where battered wives in the community went to the police and brought charges against their spouses. Such a thing would have been unheard of a generation earlier. These changes were taking place within the broader context of a society still very much dominated by patriarchal norms and ideology, so that the types of economic opportunities open to women continued to be limited by sex-typing of both rural and urban economic activities. Yet there were definite indications that women's power and status were increasing: women had been elected to the community *junta de acción comunal* (local elected groups through which rural communities could present requests for resources to the municipal councils), and a woman had been appointed mayor of Dagua in the late 1970s.

Such gains as women had achieved had been hard won, and the changes described here had been strongly resisted by the patriarchs. In 1984 there were still several observed cases of patriarchs who refused to let their wives participate in independent income-earning activities. In at least one case, however, the wife was threatening to leave her spouse as a result. Although patriarchy was still the clearly predominant ideology and form of household relations, patriarchal authority vis-à-vis both wife and child was being moderated at the household level by the changing socioeconomic environment.

Finally, the economic changes were providing numerous investment opportunities. In traditional rural economies, children generally provide the least risky and most easily accessible type of "investment" that a couple can make to provide a positive income

---

dren's work and their education. This new emphasis on education was especially significant in the case of the daughters: for the first time, many of them had an alternative to taking their places as wives in a new patriarchal unit. Many were finishing their elementary education; some were even, as we have seen, attending high school. Both mother and daughter could often be heard to express a strong preference for the daughter's education and independent employment over marriage. At the same time, both rural and urban employment opportunities for these daughters were expanding.

flow and a form of old-age security. A number of elderly men and women in El Palmar relied heavily on their youngest children, who continued to live with and care for them. The importance of these income-generating and security functions of children was being weakened, however, by the increased accessibility of alternative investments in farm modernization and technification, trade, and commerce and by improvements in capital markets and the possibility of establishing savings accounts. Some men were opting for these new investments over large family size.

The changing demographic behavior in El Palmar was a key factor in terms of the long-term prospects for the economic position of the community. A reduction in the "demographic imperative," along with better-educated offspring who would have better opportunities outside the community (as well as bringing new skills into the community), could reduce the internal dynamic of fragmentation and impoverishment. The changing relations between the patriarch and his wife and children in El Palmar were leading to reduced fertility, improved education for the children, a greater sharing of economic decision-making power, and greater innovation in farm practices. It is likely that comparable economic changes in other family-farm communities of Colombia were having similar effects on the household's internal relations and dynamic, contributing to the marked decline in Colombian rural fertility that has been observed in recent years (Ochoa 1982).[10]

While declining rural fertility will have a favorable impact on fragmentation of holdings and rural migration in the long run, it is clear that demographic pressure on the peasant land base will continue for the foreseeable future. Emigration has been and will continue to be an important means of relieving this pressure, and in this sense the pace and labor-absorptive nature of national economic development will be important determinants of the ability of rural households to establish themselves as agriculturalists. The character of national economic development will also directly influence market conditions for peasant-produced crops, and therefore the contribution of agricultural income to total household income.

10. The impact of economic development on rural fertility is an area in which much research is still needed, particularly on the demographic consequences of different crop-technology packages and market structures (exploring their effects on women's employment opportunities, for example).

## Market Conditions

The case of El Palmar has illustrated the market orientation of Colombia's small-scale agriculturalists. We have seen that this sector has historically been the principal producer of basic domestic foodstuffs. In general, market conditions for staple crops have not been propitious. This has been one of the principal factors that have kept the estates out of this sphere of agricultural production and protected it for smallholders. The market limitations derive from the fundamental structure of the Colombian political economy. While economic growth in Colombia has increased demand for foods consumed by middle- and upper-income urban households, the impact of industrial development on the market for staple crops has been more limited. Import-substitution industrialization in Colombia, as in many Third World countries, tended to be capital-intensive and therefore only weakly employment-generating. The combination of rural concentration of landholdings and limited urban-industrial employment opportunities contributed to a growing number of Colombians being dependent on purchased foods but with very limited purchasing power.

In the 1970s Colombian industry has focused increasingly on production for export (Bejarano 1978:221, 248–52). This emphasis has worsened the staple food market problem. With the export-oriented industrialization model, the need to improve domestic income levels (to broaden the industrial market base) is avoided at the same time as international competition renders labor costs more crucial. Downward pressure on wages reduces the purchasing power of the labor force and limits the market for basic foodstuffs. In 1978–79 the minimum wage in Cali of 3,000 pesos per month was barely sufficient to provide a nutritionally adequate diet for a family of four (husband, wife, and two children under ten would need over 2,300 pesos per month for this diet), leaving insufficient funds for other necessities (rent, fuel for cooking, etc.). Clearly, food expenditures for such a family would have to be severely restricted. Although in many cases urban households counted on more than one income earner, only a small fraction of working households were actually covered by the minimum wage law.

What can we conclude about future market conditions for the family farmer in Colombia? Clearly, forecasting market conditions for smallholding production is a complex problem. Prices of export products depend both on government policy and on factors outside government control. Prices of domestic products depend on overall patterns of economic development that are, again, influenced both by government policy and by international factors. A continued deterioration of international coffee prices will make it increasingly difficult for coffee producer prices to be protected through devaluation, given the inflationary impact of this policy.[11] The experience of El Palmar demonstrates clearly that the greatest stimulus to the family-farm economy is broadly based regional development, which leads to expanded markets and improved transportation facilities. The motor force for this development in Valle del Cauca has been coffee exports, based on the extensive involvement of smallholders in coffee production. Trends in international market conditions will therefore be a key factor affecting the Colombian family-farm economy, through both their direct effect on coffee producers and their indirect effect on domestic markets for other smallholder products.

Although market development is important to rural producers, it should be stressed that this development has not been sufficient to sustain the Colombian family-farm economy. Even in favored communities such as El Palmar, rural poverty and emigration have persisted. Migration, in the context of limited employment opportunities in the capital-intensive sectors of commercial agriculture and industry, has transferred the difficulties of the family-farm sector to these other regions. The problem of insufficient domestic food production has increased in the 1980s, as has the level of rural and urban guerrilla activity. These factors have led to greater emphasis on programs to "modernize" peasant agriculture as the means to increase food production and achieve social stability, in the context of the political infeasibility of land reform.

## Impact of Programs to "Modernize" Peasant Production

Phase 1 of the DRI program was considered successful in meeting its primary objective: to raise, at least in the short run, the quantity

11. On the other hand, an export-oriented industrialization strategy may make the economy less vulnerable to the negative consequences of devaluation than the

of foodstuffs produced for urban and major rural markets. As a result of this success, Phase 2 was planned in 1981 to expand the projects to eight new departments. The 1982 election of the Conservative candidate Belisario Betancur as Colombia's president after a period of Liberal rule introduced a new element of uncertainty into the DRI program. In January 1984 the program was still going through a period of evaluation and assessment, which was slowing the implementation of new projects.[12] There had been considerable turnover of DRI personnel. Nevertheless, there was a strong sense within the program that it was taking a positive new direction. This was reflected in new position papers emphasizing greater producer participation and focusing on aspects of marketing structure that had been insufficiently addressed in the Phase 1 projects (e.g., storage, transportation, and the elimination of middlemen). Along with the FDDZC program, which found similar marketing difficulties with its diversification strategy, the DRI program is now attempting to create rural cooperative organizations (*centros de acopio*) to provide these services.[13]

At the same time, the DRI program itself was moved out of its original home in the National Planning Department (Planeación Nacional) to the Ministry of Agriculture in 1983. There was some debate over whether this would strengthen or weaken the program. By the summer of 1984, proposals were circulating to create an autonomous Rural Development Institute (Instituto DRI) to focus entirely on the problems of the smallholding sector, a step that was subsequently taken (Duque 1984). Overall, it seems clear that the DRI strategy is still considered to be a viable and important one for the country, and the government can be expected to continue to support this program—within the constraints, however, of the fiscal difficulties being generally experienced by Latin American countries in the 1980s.[14]

---

import-substitution industrialization strategy, depending on the extent to which the export production is not import-intensive.

12. The changes occurring within DRI were explained to me in 1983–84 by DRI officials, as well as by other Colombians who had been involved in evaluating the DRI program and developing proposals for the new projects.

13. The emphasis of the FDDZC program was detailed in a newspaper article (*El Siglo* 1983) and discussed with me by several Coffee Federation officials. The new emphasis of the DRI program was described to me in interviews with DRI personnel and is summarized in *El Siglo* 1984.

14. In Colombia the early 1980s was marked by a sharp decline in overall rates

That being the case, it is important to consider the effectiveness of the DRI program in addressing the problems of the smallholding sector. We saw that the DRI program was designed specifically to overcome the historic lack of access of the small-farm sector in Colombia to commercial inputs (and credit to finance their purchase), and to complementary infrastructure such as water and transportation. These are clearly necessary inputs for the family-farm sector. The new emphasis on cooperative marketing structures may also be beneficial to this sector.[15]

Nevertheless, I find it questionable whether the DRI approach alone will meet the needs of the family-farm sector. First, the program has dropped large segments of the rural population as potential beneficiaries: following the recommendations of the Phase 1 evaluations, Phase 2 concentrates on households with "adequate" landholdings. The program is clearly becoming primarily a production strategy, and a short-term one at that.[16] Its principal objective is to reinforce the production role of the family-farm sector within the overall economy, rather than to improve the income position of the small-farm households as a whole. The case of El Palmar suggests that it is possible to devise crop-technology packages that can improve income for households with very small holdings (or even for households whose only access to land is through sharecropping or rental arrangements, depending on the conditions of those arrangements). This is, however, more problematic than focusing on the larger family farms and is not likely to become part of the DRI

---

of economic growth. While the coffee boom of the 1970s contributed to an annual average GDP growth rate of 6.4 percent in that decade, the international recession of the early 1980s reduced the rate of GDP growth to 2.5 percent in 1981 and to 1.5 percent in 1982 (World Bank 1979: 128, table 2; Fajardo 1983: 152).

15. A recent study of the marketing of perishable foodstuffs in the Bogotá area, however, has concluded that the operations of market intermediaries are not the principal cause of low farm income for smallholders (Forero and Rudas 1983).

16. This emerges clearly from the discussion of the elimination of some DRI projects in the December 1982 document (Dirección General DRI–PAN 1982a), where we find statements that "the optimization of the operation of DRI requires a concentration in areas with a recognized production of marketable surpluses in order to avoid the costly dispersion of resources," and that in light of "the need to concentrate state investment in areas with the greatest productive potential . . . it has been considered desirable to give priority to those regions that, from the point of view of the DRI objectives of obtaining an important increment in the supply of foodstuffs in the short term, have the necessary and sufficient characteristics to achieve this goal" (translated from pp. 12–13).

program in the near future. Land-poor and landless households will only receive such possible indirect effects of the DRI program as new employment opportunities.

Even assuming a wider population could be targeted, there are fundamental weaknesses with the peasant modernization strategy's ability to achieve its dual goals of increased staple food production and reduced rural poverty. The market conditions that discourage estate production of basic foodstuffs also render those crops risky ventures for a small-farm modernization strategy involving the increased use of purchased inputs. For this reason, the success of the economic modernization of El Palmar was predicated not on the increased cultivation of staple crops, but rather on a replication of the strategy of the large estates: production for the more lucrative export and urban higher-income markets. The clear trend in El Palmar over the 1970s was a shift on the small farms away from traditional staples such as maize, beans, and root crops, towards crops such as tomatoes and coffee (with plantains) for these other markets, given the much higher return *per unit of land* from the new crops and varieties.

The trend observed in El Palmar was not an isolated instance. A comparable shift of smallholding production away from such basic foodstuffs as maize and cassava towards the cultivation of coffee, tomatoes, and pineapples was found by Roldán (1979) in his interviews throughout the mountain zones of western Valle del Cauca. This is related to the fact that the new yield-increasing technology available to these mountain farmers was principally geared to the nontraditional crops and varieties. The Coffee Federation's FDDZC program, an important influence in this area, has emphasized products "for industry, import-substitution, or export" (*El Siglo*, 11 November 1983:9). This leaves open the question of which sector will step in to maintain production of domestic staples, and what will happen to rural and urban low-income households as staple food prices rise.[17]

Finally, while modernization can increase rural income, the income gains per household are likely to be limited, as we have seen in the case of El Palmar. Some of the technical difficulties en-

17. Without those price increases being sufficient, however, to stimulate staple food production in the context of a diminishing land base and limited technological change for those crops.

countered with the modernization programs could theoretically be corrected through the development of more appropriate crop-technology packages. Some groups in Colombia working to improve the integrated rural development strategy are emphasizing this approach.[18] The DRI and FDDZC programs could move in the direction of small-farm agricultural "systems" that involve a variety of products and emphasize soil preservation. These modifications, however, emulate the responses of the peasant producers themselves in trading income gains for income security. Fundamentally, given the limited income gains that can be obtained per acre, the peasant modernization strategy fails to address the basic obstacle to long-term rural prosperity in Colombia: the shortage of land in the family-farm sector.

The DRI approach stems from an essentially Chayanovian analysis of the family-farm economy that emphasizes vertical concentration as the family farmer's problem, while overlooking the more basic issue of access to land. Although complementary resources and improved market conditions have increased the contribution of agricultural income to the livelihood of many rural households, they have not created the conditions for the intergenerational reconstitution of those households as full-time family farmers. This can only be accomplished through land reform or a strongly employment-generating national development strategy that draws large numbers of people out of the rural sector. As these are both unlikely in the near future, the conversion of the "semiproletarian" households as a group into full-time agriculturalists will not occur. Market expansion and technological innovation will be insufficient, in the context of the overall scarcity of employment opportunities outside the peasant sector and of land in that sector, to enable peasant households on average to become full-time family farmers.

This being the case, an important question is whether peasant

18. For example, FUNDAEC (the Foundation for the Application and Teaching of the Sciences), based in Valle, has been working to develop subsystems of mixed crops that can be grown on small plots. By late 1983 they had identified fifteen subsystems involving various combinations of basic foodstuffs (beans, maize, cassava, etc.), fruits, and livestock (cows, pigs, chickens), which they had found in trials to be technically feasible and acceptable to the producers. FUNDAEC is working on its own and in conjunction with other groups in Valle del Cauca to present its findings to peasant producers. (Based on internal FUNDAEC documents and an interview with the acting director of FUNDAEC in January 1984.)

households will even be able to maintain their status as part-time family farmers over their life cycles. This depends on future trends in the factors that have enabled small-scale commodity production to persist in Colombia. What will be the impact of continued economic development on the internal relations of the household production unit and on the technical conditions of agricultural production?

## Economic Development, Technological Change, and the Competitiveness of the Family Farm

It is clear that technological changes are taking place in Colombian agriculture, both in the estate sector and on the family farms themselves. As we have seen, there has been considerable debate regarding the impact of technological change on the family-farm economy. At the heart of this debate lies the question of the effect of technological change on the competitive position of the family farm versus large-scale agricultural production.

### *The Family Labor Process*

There is substantial evidence from Latin America that the traditional family farm has much greater land productivity (higher yield per acre) and lower labor productivity (output per hour of work) than the traditional estates (Barraclough 1973). This type of evidence has been used to support the position that the traditional family farm maintains its competitive position in the output market by "super-exploiting" family labor—that is, by continuing to expend labor time far beyond the point at which the marginal product of labor equals the going wage rate in the agricultural labor market. The traditional family farm is therefore able to offer products for sale in the market at prices below those of large-scale units dependent on market labor, because the family farm is "devaluing" the family labor input.[19]

Such seemingly irrational behavior becomes rational when the

19. Banaji (1977:32–36) provides a good elaboration—and critique—of this position; see also Amin 1974; Servolin 1972; and Corchuelo 1979.

market wage really does not reflect the opportunity cost of family labor—where, in other words, there is some impediment to off-farm employment. This may involve, for example, the evening or weekend work of male family members on the farm (i.e., during periods of "leisure" time). Or it may involve the ability to use female and child labor on the farm in circumstances where their opportunity cost is zero (because farm work is compatible with household work or because of social impediments to employment opportunities for children or females off-farm). Of particular importance here is the ability of the patriarch to command the on-farm labor inputs of household members.

As we have seen in the case of El Palmar, however, the advantage to the family farm of the ability to "super-exploit" family labor may diminish with economic development that expands the reach of the labor market, increases the opportunity cost of family labor, and weakens the patriarch's authority over that labor. As subordinate household members begin to find local employment opportunities, the market opportunity cost of their time must be considered in allocating farm tasks. This is particularly clear once the patriarch has to pay his sons the going wage for their labor input. The family labor advantage also diminishes as the farm becomes more specialized and therefore more dependent on hired labor for peak-demand tasks, as was the case with the coffee harvest in El Palmar. Although there may still be a considerable amount of family labor that has zero opportunity cost (the labor of children after school and on weekends and holidays, the evening and weekend labor of adults, the labor of women while caring for young children), it is clear that economic development can make significant inroads into the labor-cost advantage of the family farm. As chapter 6 demonstrates, the new economic opportunities can pull family members in different directions, fragmenting the family labor process and weakening the patriarch's ability to command household labor for agricultural operations.

A related argument with respect to the competitive position of the traditional family farm holds that its competitiveness derives from the fact that the market price does not have to cover rent, interest, or profit (since the family farmer is at one and the same time landowner, banker, and entrepreneur). Once family farmers begin to capitalize their farms, however—borrowing for the purchase of land, equipment, and other farm improvements—their returns do have to cover the opportunity cost of capital. This was increasingly the case in El Palmar. Thus again, economic development and modernization may erode this cost advantage of the family farm.

Ultimately, as the economic environment of the family farm becomes more capitalist, in the sense that markets become more perfect and the opportunity cost of all factors of production is increasingly set in those markets, the behavioral characteristics frequently pointed to as the source of the cost advantage of the traditional family farm decline in importance. If it is assumed that these behavioral characteristics are the sole factor allowing the family farm to compete with large-scale units, the inevitable conclusion from these observations is that family-farm production is doomed to extinction by processes of economic development and agricultural modernization.

### *Technical Cost Advantages*

Yet we have seen that these very processes, far from dissolving the economy of El Palmar, gave it a new lease on life. Although behavioral characteristics may indeed have been important to the traditional family farm, the experience of El Palmar suggests that the economic competitiveness of the family farm derives from more than just its ability to ignore certain costs of production. The erosion of the behavioral differences between the family farm and large-scale production units reveals more clearly that a possible underlying source of family-farm competitiveness is a technical cost advantage deriving from the lack of scale economies, or even the existence of diseconomies of scale, in some branches of agricultural production. This may be a factor in the competitiveness of the traditional family farm versus the traditional estate, although the behavioral cost advantages may be the most significant in this case. The technical, as opposed to behavioral, cost advantages become more important when comparing the traditional family farm to the modern large-scale production unit; and they become central once the family farm modernizes and therefore sheds much of its behavioral cost advantage.

A particular "division of labor" has been evident for some time in Colombian agricultural production, with traditional small-scale farm units being the principal producers of staple food crops, the large traditional estates emphasizing cattle production combined with the rental of land to small-scale cultivators, the large commercial estates producing export crops and increasingly industrial inputs, and farms of all sizes producing coffee. This product and market division can be explained by the technical characteristics of these different branches of agricultural production as influenced by agricultural research and by government policy. With respect to tropical staple food production, various studies, which the research in El Palmar supports, indicate that under the existing state of technological development there are few or no technical economies of scale associated with the production of many of those crops. As a result, the estates have left these crops to the traditional smallholding units (USDA 1970: 23),[20] who continue therefore to exercise an important role as agricultural commodity producers. The continuing economic competitiveness of family farmers in some spheres of agricultural production underlies the division of labor in Colombian agriculture.

This division of labor is not necessarily permanent, since agricultural research may be able to alter the technical conditions of production. In recent years, for example, there have been efforts to develop new technology that will render tropical products more amenable to large-scale cultivation. In Colombia, the research center CIAT (Centro Internacional de Agricultura Tropical) is in the vanguard of such efforts, currently working on cassava production as well as on other tropical products. As this research proceeds, the

20. Rice is the exception: irrigated rice cultivation, which was developed on the large estates controlling the lowlands, has a much higher yield than drylands rice, grown on the smallholdings of the mountain regions (USDA 1970: 23). In the case of corn, production has been adopted on a large scale by Cauca Valley estates, using improved varieties and modern technology. However, studies indicate that this is not adversely affecting smallholding maize production, as the estates and the smallholders are producing for different markets. While the estates concentrate on feed corn, the smallholders are emphasizing traditional maize varieties that are staples in rural and urban diets (Torres 1968). A similar situation is developing with sugar cane, an important crop of the family farmers throughout Colombia's southwestern mountains: the mountain farmers process their cane into *panela*, which is sold principally in the mountain zone itself, while cane grown on Cauca Valley estates is processed primarily into refined sugar and secondarily into *panela* for the major urban markets of the valley (Roldán 1979: 194–98).

obstacles posed by the technical conditions of production to the extension of estate cultivation into the sphere of tropical foodstuffs may well be eroded. The specialized varieties and new technical processes that emerge from this research may give technical cost advantages to large-scale production.[21] As this occurs, the ability of the traditional family farm to compete in these markets will depend solely on the behavioral cost advantages discussed above, and the strength of those advantages compared to the new technical cost disadvantages.

Increasingly, however, the traditional sectoral division of labor in Colombian agriculture may be transformed not so much by the incursion of modern estates into the production of traditional staples as through the modernization of the family farms and their movement *out* of traditional staples and into export crops and nontraditional foodstuffs. This movement has been facilitated by the introduction of new technology into the family-farm economy itself. The key issue with regard to the continued production role of the family-farm sector is therefore becoming the relative competitive position, not of the traditional family farm vis-à-vis either the traditional or the modern estate (the two comparisons most frequently found in the literature), but rather of the modern family farm vis-à-vis the modern estate.

It has been argued that the family farm is an inefficient unit of production because it is inherently incompatible with technological change, particularly mechanization (Banaji 1977:36). There are three weaknesses of this argument. First, mechanization per se may be unrelated to economic efficiency, as measured in terms of total costs per unit of output. While mechanization can increase the productivity of labor, it does not automatically increase total factor productivity, and therefore unmechanized small farms are not necessarily inefficient. Second, it is not clear in any case that the family farm is incompatible with mechanization. Mechaniza-

21. A summary description of some of CIAT's current research efforts in staple food crops recently appeared in the *New York Times* (Crittenden 1981). The conclusion drawn here about these efforts is based on observations of CIAT programs and conversations with CIAT personnel, rather than on any explicit statement of purpose by CIAT itself. Certainly some of the CIAT research does have considerable potential usefulness for small farmers. The question is whether, as with the Green Revolution technology, it will have even greater usefulness for large-scale producers.

tion can be a means of strengthening the family farm by obviating the need for hired labor during periods of peak labor demand: the use of tractors and harvesting equipment by U.S. family farmers is a case in point.[22]

Third, technological change can take forms other than mechanization, and, as has been demonstrated in El Palmar, new yield-increasing technology can be quite compatible with the family-farm sector. If we consider the food crops increasingly being grown in El Palmar, we find that the yield-increasing technology associated with the cultivation of these crops appears to have intensified, rather than eroded, the technical cost advantage of small-scale producers. Production data for El Palmar reveal diseconomies of scale in the cultivation of tomatoes and green peppers. These diseconomies were related to the extreme sensitivity of these crops to tropical pests and diseases, resulting in the need for constant care and supervision. The genetic weakness of many of the new hybrid varieties and their consequent susceptibility to disease increases the competitive position of the small cultivator. This explains why much of the production of nontraditional fruits and vegetables in Latin America is being carried out by small-scale cultivators on contract with processors and distributors, even in cases where the cultivated land is actually owned as a large unit. These are various forms of the "vertical concentration" of family-farm agriculture postulated by the Chayanovians.[23] Both the DRI programs and the FDDZC programs in Colombia have emphasized crops with limited scale economies and rapidly encountered diseconomies, particularly fruits, in addition to the vegetables we have discussed in the case of El Palmar.

With respect to coffee, there has recently been considerable debate in Colombia over the effects of "technified" coffee cultivation on the competitiveness of the small producer. It has been argued that the smaller farms are being driven out of coffee production by these changes. The case of El Palmar suggests, however, that this is

22. This argument is developed in detail by Brewster (1950). Of course, the efficient use of farm equipment may require the physical expansion of the family-farm unit and therefore a decreasing number of family farms, as has been the case in the United States.

23. For examples see NACLA 1978 on Carnation in Peru and Taussig 1974 on sugar production in Colombia. Contract farming is also an increasing phenomenon in the United States (Raup 1973; Science for the People 1979).

not occurring uniformly. Much of the new coffee technology, including the new varieties of coffee that are being developed, has been easily incorporated into the family-farm system of production and has not given the larger units a clear competitive edge over the family farms. Again, the new technology has not sufficiently reduced the inherent obstacles that agricultural production poses to large-scale cultivation. Moreover, it appears that the large coffee farms are more sensitive to price changes than the family farms: while most of the absentee owners of modern coffee farms interviewed in El Palmar said that they would sell out if coffee prices fell, none of the family farmers gave that response.[24]

The implications of the foregoing analysis for the Colombian family farm, and in general for the resolution of the "peasant question" in Latin America, are the following. First, the behavioral cost advantages of the traditional family farm are eroded as it becomes more highly capitalized and as off-farm economic opportunities for subordinate family members increase. These tendencies are already visible in El Palmar and are likely to be intensified throughout Colombia by the country's continuing economic development and by government and private rural development programs. However, the family farmers of El Palmar have continued to find areas of production in which they are competitive with large-scale production. This is because of the fundamental advantage of the family-farm organization of production, which lies not in its ability to "super-exploit" family labor, but rather in the obstacles that the natural characteristics of many branches of agriculture pose to the large-scale organization of production. A great danger to the family farm is therefore the development of agricultural technology that erodes these obstacles, increasing scale economies and reducing scale diseconomies. This process may attract large-scale units of production increasingly into the spheres of agricultural production that have heretofore been the preserve of the family farm.

24. For a good assessment of the overall position of the small coffee producer in Colombia, see Palacios 1980:241–47. Where small coffee producers have been displaced, they may well have been sharecroppers adversely affected by the land legislation of the 1960s (Law 135 of 1961, Law 1 of 1968), as were sharecroppers in general (Perez Zapata 1978). Owner-operators, as in El Palmar, have been in a position to benefit from some components of the new technology under the price umbrella of the Coffee Federation, while being less affected than large producers by rising input costs (because of greater reliance on nonpurchased inputs, such as solar energy for coffee drying).

### *Production Flexibility*

Increased external competition is not the only danger, however, to which technological change may expose the family-farm economy. There is another potential weakness that technological change may introduce. The greatest strength of the family farm, as we have seen clearly in El Palmar, is its flexibility—its ability to reallocate resources quickly in response to changing market conditions. That is because its resources have been principally the malleable ones of land and labor. "Modern" technology may, however, involve the increased introduction of fixed capital into the family-farm operation, particularly in the form of crop-specific equipment and structures. This is becoming increasingly the case in El Palmar, where we have observed households becoming indebted to finance the installation of costly equipment to process and dry coffee. The new coffee trees themselves now represent considerable investment, often debt-financed; another recent example is the extremely large debt incurred by one farmer to plant pineapples.

This capitalization of the family-farm production process clearly reduces its flexibility, locking the household into specific crops even in the case of falling relative prices. Increased competition or deteriorating market conditions might then threaten the viability of the family-farm unit, resulting not in economic reorganization of production but rather in farm loss. To avoid this development, research is needed on technology that increases rather than decreases the flexibility of the family-farm unit. To date, little emphasis has been placed on this type of research.

## A Synthesis

How has peasant commodity production been able to persist in Latin America? I have argued here that the technical conditions of agricultural production provide a key part of the answer. The natural characteristics of agricultural production are such as to pose obstacles to the development of capitalist agricultural production. In particular, the sequential character of agricultural production and the managerial diseconomies arising from its biological and spatial character favor small-scale production, while the seasonality of labor requirements may give an advantage to mixed-crop, family-labor-based agricultural systems. Differences in these technical characteristics between agricultural activities leave scope, however, for the coexistence of a variety of forms of agricultural production.

Nevertheless, these technical characteristics create only the physical parameters for agrarian structure in any concrete case. Social factors are also an important determinant of agrarian structure. Economically competitive forms of production may, for example, be forcibly dispossessed. The state may intervene with price or support policies, including support for technological research, that favor one or another producer group despite its intrinsic lack of economic competitiveness.[25] Because of both the technical and social factors, the development of agrarian structure is problematic rather than predetermined and inevitable.

The methodological implication of this theoretical perspective is that the analysis of agrarian structure should depart from an examination of these technical and social factors in each concrete case in order to understand the fundamental conditions of existence of each productive form and the manner in which those conditions have been secured. This necessarily entails an examination of the nature of the state in each specific instance: the coalition of forces represented at the level of the state, the pressures and constraints on state action, and the effect of state policy on the conditions of existence of the contending organizational forms. It also, however, entails an analysis at the level of the individual farm unit to determine, in the first instance, the nature of that unit and its conditions of existence. This is the theoretical and methodological framework that I have applied to the analysis of the development of the Colombian family farm.

Overall, these farmers have been protected from direct competition from larger units by the unstable market conditions and low product prices they face for some products, while being able to compete effectively in other markets because of the limited development of technical economies of scale. The poor quality of the land they farm has also protected mountain farmers from direct encroach-

25. Servolin (1972), for example, has reported the conclusions of a group of French scholars that it was government support of agricultural prices that made possible the development of capitalist agriculture in that country. In the United States, Raup (1973) has argued that the recent expansion of corporate farming had its roots in tax policy and government farm-support programs.

ment by larger units. Furthermore, the topography of their land poses obstacles to the mechanization of many farm operations, and therefore to competition for that land from large-scale producers.

All of these characteristics, while discouraging competition from larger units, have also posed serious difficulties for the family farmers themselves. Today these farmers are facing a crucial stage in their development. Having expanded in the mountain regions to the limits of the available land base, with the majority of Colombia's readily accessible land held by the large estates, the family farmers have been experiencing an intensified process of fragmentation of holdings and soil deterioration. While the reduction of the demographic growth dynamic of the family-farm economy will have positive implications for the ultimate reduction of pressures towards fragmentation and soil exhaustion, these pressures are clearly going to continue for the foreseeable future. Colombia's peasant producers are increasingly turning, where possible, to nontraditional crops and technology that can increase yield and, potentially, income per acre.

Our analysis of Colombia's political economy indicates that this process of family-farm modernization is meeting important needs of the country as a whole. It can increase the output, and thereby reduce the prices, of foods for urban markets; it can increase production for export and therefore the foreign exchange needed to support industrial development; it can provide markets for domestic products, both agricultural inputs[26] and consumer goods purchased by the more prosperous rural community.[27]

26. In El Palmar, for example, we have seen the increased purchase of such domestically produced inputs as *gallinaza*, *caturra* seeds and seedlings, and coffee-processing equipment, as well as the stimulus provided to the local construction industry and its suppliers. In addition, and perhaps not so positively for the economic development of the country as a whole, the farmers of El Palmar have become increasingly dependent on imported inputs such as agricultural chemicals. Farmers in El Palmar were observed to be using a variety of imported pesticides such as Aldrin and Dieldrin, which have been banned for use in the United States because of their danger to human health, without any precautionary measures.

27. Many economists have argued for these virtues of the competitive family-farm structure in developing countries, pointing to its efficiency and contribution to overall economic development. Owen (1966) has presented the classic argument for the family-farm structure in developing countries in terms of its receptivity to technical change and its tendency to increase production. Johnston and

In the context of growing urban difficulties related to the stagnation of domestic food production and rapid rates of rural-urban migration, the government of Colombia has attempted to encourage this small-farm modernization process through its integrated rural development program, Programa DRI. Other organizations in Colombia, such as the Colombian Coffee Federation, have also been working with small farmers to alter their crop-technology choices. These efforts are likely to continue, within the fiscal constraints imposed by international market conditions. The Colombian family farmer can be expected to continue to adopt new crops and technology in the context of continued fragmentation of holdings and soil deterioration, with production geared increasingly towards export and urban high-income markets.

This strategy for the family farm has led, as in the case of El Palmar, to income gains and to demographic changes favorable to the long-term health of the family-farm economy. It has meant that farm income continues to make an important contribution to the livelihood of many rural households. While market expansion and technological innovation have benefited the family farm, however, the nation's overall development has generated economic, political, and ideological challenges to the patriarchal household relations that traditionally ensured a subordinate, efficient family labor force. Furthermore, the new technology being adopted by the family farm may ultimately limit its flexibility and expose the family farm to increasingly difficult competition from large-scale production.

Whether improvement of market and technical conditions strengthens or weakens the family farm will depend on the type of technology developed and on the social and political conditions under which those improvements are brought about. There are clearly strong pressures operating in the direction of the development of technology that reduces the competitiveness of small-scale production. These pressures include the weight of imported technology and research developed in a labor-scarce environment, the political influence of large landowners, and the tremendous resource costs involved in creating basic infrastructure for a modern family-farm sector. There is in many circles, on the other hand, a

Kilby (1975), among others, have argued for the family-farm structure in terms of its contribution to market growth as well as its productivity.

recognition of the social and political costs of such a large-scale agricultural development strategy, given the sheer weight of the rural population and the inability to absorb that population in other productive activities. Whether that recognition, expressed in the integrated rural development strategy, will be translated into effective policy in the face of the forces arrayed in favor of the large-scale strategy of agricultural development remains to be seen.

In any case, the types of technological developments that may prove harmful to the family farm's competitive position vis-à-vis large-scale production will, if they do occur, be some time in coming. We can safely assume that the family farm will continue for the near future to find crops and markets in which it can successfully operate, although the farm households will continue to have to combine farm production with other types of economic activities in order to assure their livelihoods. Unless mounting social tensions lead to a radical transformation of Colombian society, rural households will persist as part-time family farmers for many years to come.

# Appendix A: Population Growth in Valle del Cauca and Colombia

Table A.1 summarizes the available data on population growth in Colombia, Valle del Cauca, Cali, and the western mountain region from the late eighteenth century to the early 1970s. This appendix discusses the data for Valle del Cauca in the nineteenth century, when frequent redefinitions of political boundaries must be taken into consideration.

After independence from Spain in 1820, the area of Valle was divided between two provinces: Buenaventura, which included the political subdivisions (*cantones*) of Cali, Roldanillo, Raposo, Iscuande, and Micay; and the province of Popayán to the south, which also included the rest of the Cauca Valley land. According to the 1825 census, the population of the province of Buenaventura was at least 17,684. Urrutia (1970: 18) points out that the census underrepresents the population of this period, so that this figure should be considered a minimum.

Buenaventura was the province with the highest percentage of slave population in Colombia at this time: 38 percent. The next highest province was the Chocó, with 28 percent. According to Vergara y Velasco (1974: 925), 25 percent of the population of the Cauca region as a whole (including the Chocó) were blacks in 1810. This was a higher percentage than in any other region of the country. The blacks in Cauca accounted for 42 percent of all blacks in Colombia at that time, reflecting the greater role that slavery played in the southwest than in the rest of the country.

The population of the province grew rapidly, with an average annual increase of 2 percent between 1835 and 1843, slightly above the 1.8 percent average for the country as a whole (growth rates are calculated from the data in table A.1). Census figures for 1835 and 1843 show that the

TABLE A.1. *Colombian Population Data, 1787–1973*

| | Colombia | Valle del Cauca[a] | Cauca (and Chocó)[b] | Province of Buenaventura[c] | City of Cali | Municipio of Dagua[d] |
|---|---|---|---|---|---|---|
| 1787 | — | — | — | — | 5,690* | — |
| 1825 | 1,228,259 | — | — | 17,684 | — | — |
| 1835 | 1,686,038 | — | — | 31,920 | — | — |
| 1843 | 1,931,684 | — | 268,607 | 37,104 | — | — |
| 1851 | 2,175,459 | — | 319,898 | — | — | — |
| 1864 | 2,662,812 | — | 386,208 | — | — | — |
| 1870 | 2,931,984 | — | 435,078 | — | — | — |
| 1905 | 4,488,782 | — | 539,672 | — | — | 2,331 |
| 1912 | 5,072,604† | 217,159 | — | — | 27,700‡ | 3,146 |
| 1918 | 5,855,077† | 260,225‡ | — | — | 45,600‡ | — |
| 1928 | — | 430,739‡ | — | — | 59,500‡ | — |
| 1938 | 8,701,816† | 613,230‡§ | — | — | 88,366‡§ | 14,149§ |
| 1951 | 11,548,172 | 1,106,927‡ | — | — | 243,600‡ | — |
| 1958 | — | — | — | — | 464,800‡ | — |
| 1964 | 17,484,508 | 1,733,053 | — | — | 637,929 | 25,508‖ |
| 1973 | 22,551,811 | 2,353,284 | — | — | 967,908 | — |

region's slave population declined steadily in relative importance during this period. It is difficult after this point to reconstruct precise figures for the region because of frequent redefinitions of political jurisdiction in the nineteenth century. In 1835 the province of Cauca was created out of the Cauca Valley land that had been part of the province of Popayán. In 1851 the state of Cauca was created, incorporating all the lands of the Chocó, Buenaventura, Popayán, and Cauca provinces.

Because of these and subsequent shifts, the best we can do is to look at population growth for the southwestern region as a whole. Even with this data, however, there are census errors related to boundary changes as well as to incomplete enumeration (Vergara y Velasco 1974:937–41). The newly created state of Cauca, with approximately 268,607 inhabitants in 1843, grew at an average annual rate of 2.1 percent between 1843 and 1864, approximately the same as the earlier rate for Buenaventura. The population of Cauca continued to grow at the same rate for the next six

*Sources:* The principal sources were the Colombian Population Censuses. Those for 1825, 1835, 1843, 1851, 1864, 1870, 1905, and 1912 were consulted in a bound, untitled volume in the Population Census Office of DANE (Departamento Administrativo Nacional de Estadística) in Cali. Those for 1951, 1964, and 1973 are reported in United Nations 1955: Tables 1 and 5; 1972: Tables 2, 8, and 9; 1980: Tables 3, 8, and 30. Reported figures for these three censuses vary; the most recent for each census is presumably the most reliable. See also DANE 1973b. Census data for 1825 through 1870 is also reproduced in Gomez (1970), along with a discussion of the reliability of this data. Gomez concludes that while the figures for 1835, 1843, 1851, and 1870 appear to be reasonably accurate, the 1825 and 1864 figures are highly questionable. There are some discrepancies between the figures in the DANE census volume and in the Gomez article.

Other sources:

*DANE 1955.

†United Nations 1949:91.

‡Camacho Peréa 1962:92–94.

§DANE 1942.

‖DANE 1975; also in Instituto Geográfico Agustín Codazzi 1971.

[a] Prior to 1910, Valle del Cauca did not constitute a separate political entity. The population of what was to become Valle del Cauca was included variously in the provinces of Buenaventura, Chocó, Cauca, and Popayán (see, for example, Diaz del Castillo 1937:53).

[b] The state of Cauca was created in 1851 out of parts of the old provinces of Buenaventura, Cauca, Popayán, and Chocó. Before the 1912 population census a redefinition of political boundaries led to the splitting off of the *departamento* of Valle del Cauca. The southern area, which remained under the jurisdiction of the *departamento* of Cauca, and the northern area, which became the *departamento* of Chocó, are not covered in this study. The data presented here for Cauca from 1843 to 1905 includes the population of the original Chocó province, some of which was ultimately joined to the *departamento* of Valle del Cauca.

[c] After 1843 the province of Buenaventura is included in Cauca and later Valle del Cauca.

[d] Officially created in 1868. Until 1848 the area was known as the Parroquia de San Juan del Salado; between 1848 and 1868 variously as Juntas de Dagua, Papagalleros, and El Queremal. Between 1868 and 1904 the district was renamed several times but kept its physical boundaries (Becerra Collazos 1977:6; Banderas 1944:111).

---

years, reaching 435,078 by 1870. The population of Colombia between 1843 and 1870 grew at an annual average rate of 1.9 percent, so that Cauca was continuing to grow at just above the national rate in this period.

No more population data are available until the early 1900s. By then the population of the states of Cauca and Chocó was 539,672—an average annual rate of growth of only .7 percent between 1870 and 1905. During this same period the population of the country as a whole grew at an average rate of 1.5 percent, and some states were growing even more rapidly, with Antioquia experiencing a 2.4 percent average annual growth rate. Thus after an early rapid rate of growth for the province of Buenaventura and subsequently for the new state of Cauca, the rate of population growth for the southwest fell to considerably below the national average by the 1870s.

However, it is difficult to draw conclusions about the Valle region from the Cauca figures, since in addition to Valle those figures included the regions of the Chocó mining district and the lands to the south of Valle. The economies of both these areas were stagnating, if not actually declining, in the second half of the nineteenth century, reflected in the decline and finally the abolition of slavery, on which the mining activity of this region was highly dependent. An indication of this decline is to be found in the sharp decrease in the slave population of the region both as a percentage of the total population and in absolute terms. In 1825, as noted previously, 37.8 percent of the population of Buenaventura, or 6,690 persons, were slaves, and in the Chocó province 28.1 percent, or 4,843 persons, for a total of 11,533 slaves in the two provinces. In 1835 the provinces of Chocó, Buenaventura, and Cauca had a total of 13,271 slaves, which amounted to only 12.8 percent of the total population. By 1843 the numbers for these three provinces had declined to 9,385 slaves, or 7.5 percent of their combined population. In the province of Popayán a comparable decrease is evident (figures from the 1825, 1835, and 1843 population censuses). As Fajardo (1983) points out, this decline in the slave mining population, much of which took place through manumission (formalized by the manumission law of 1821), was reflected in a growing population of free black settlers.

Within this overall southwestern region, Valle was likely to have been the principal growth center. This is true not only for the late nineteenth century, when the Antioquia migration began to reach into the northern mountains of Valle, but also from the middle of the century, when we see the steady development of the mountain region between Cali and Buenaventura.

# Appendix B: The Household Interviews

Fieldwork was carried out in El Palmar from October 1978 to June 1979, following a preliminary investigation of southern Valle agriculture in April–May 1978 and a period of more intensive study of background information on the El Palmar region during September of that year. Armed with what turned out to be misleading maps and dated, but somewhat more accurate, population census information (DANE 1973b; original returns consulted in the Cali office of DANE), I moved into a room in the El Palmar schoolhouse as soon as the school year began. I lived in the community for a month before beginning the formal interviews. This was important not only to get a feel for the community's history and for those factors that appeared to have been central to its development, but also to introduce myself to members of the community and explain the purpose of my presence. I also needed to determine the physical boundaries of the community, given the lack of accurate detailed cartography of the area. During this initial period, I tested my preliminary interview format on one household and eliminated or revised inappropriate or confusing questions.

Formal interviews were conducted with the head of household and/or the spouse in 105 resident households and with nine absentee landowners or their farm managers between November 1978 and June 1979. The interviews included detailed questions on the current economic activities of each member of the household, on all land transactions (purchases, inheritances, sales, rentals) that had been entered into by the household, and all aspects of current farm production. The interviews varied in length from approximately one to two and a half hours, depending on the size of the family, the complexity of the agricultural operations, and the number of land transactions to be recorded (copies of the interview sheets are reproduced in Reinhardt 1981). As the interviews were conducted, the format was periodically reexamined, and questions that were obtaining a poor response rate or were judged to be of secondary importance were eliminated. Information on some questions is therefore not available for all households.

For every interview that was conducted, there was an additional one to two hours' work checking and filling in some of the codes, calculating income and farm costs from the raw data obtained in the interviews, and checking for inconsistencies. A careful consistency check was necessary because of the sensitive nature of some of the questions and the difficulty of reconstructing some of the requested information. The first check compared the annual expenditures on foodstuffs, based on the weekly average given, to the annual income of the household as calculated from the data given on farm operations, on nonfarm earned income, and on loans and aid. A second check compared calculated income, food outlays, and other information obtained in the interviews on the economic position of the household with my informal observations and overall impression of the appearance of the household, the farm, and the family members themselves. Although some interviews were conducted by an assistant, Fernando Uribe, I generally conducted a follow-up visit to familiarize myself personally with the household. A third check compared information obtained in the interviews with information obtained from the original returns of the 1978 Agricultural Census of Valle del Cauca (CENAGRO 1978) and the 1973 Population Census (DANE 1973b).

If there appeared to be any inconsistencies, I consulted a local informant. This individual was well acquainted with all community members, and she or her family could frequently explain the source of the inconsistency. Occasionally it was necessary to speak to the household a second time in order to clear up remaining discrepancies.

## Coverage of Interviews: Households Excluded from the Sample

The 114 interviews do not represent a complete survey of households or landholdings in El Palmar. Because the purpose of the investigation was to examine the impact of economic development on community residents, I did not interview migrant agricultural workers and some roadside residents.

The migrant agricultural workers are primarily young, single men, who come to El Palmar mainly during the coffee harvest, remain a few months, and then return home or move on. They are employed principally by the larger absentee landowners, where they are housed in farmhouse dormitories. From my conversations with workers and employers in El Palmar and with agricultural extension agents in Cauca and in Nariño, I established that these young men come principally from the temperate mountain regions of Cauca and Nariño, where their families generally have small plots of land. Many of them return to their homes after the harvest. They therefore constitute a semiproletarian labor force, whose sustenance during part of the year derives from their participation in their families' agricultural operations.

The farm manager or overseer in El Palmar, by contrast, is generally older, and may have one or more rooms of the farmhouse for himself and his family. His wife does the cooking for the migrant farmhands—generally, although not always, without pay. These farm managers, called *mayordomos* or *agregados*, are either from resident families with little or no land (who may live in their own nearby house), or are brought in from the outside. In the latter case, they tend to remain for longer periods of time than the migrant laborers. For this reason, all farm managers were included in the formal interviews.

The roadside residents not interviewed were recent, landless immigrants falling into one of two categories. The first consists of extremely poor families who are "drifters." They occupy existing shacks by the roadside, often without the permission of the landowner. They may obtain work as agricultural day laborers, wash clothes, or perhaps sell local produce to passersby. They have come to El Palmar because that is where the highway took them; they form few connections with the community, do not send their children to the elementary school because they cannot afford clothing or the matriculation fees, are not known by name by the residents of El Palmar, and generally move on after a short time.

The second group of roadside residents consists of recent immigrants who arrived with capital or some trade and have set themselves up in business, catering primarily to nonlocal customers. There are two families with roadside restaurants and two butchers. There is a dentist who rents a house by the roadside and works in El Treinta. There is a police station occupied by one policeman and his family. The other policeman rents a small house jointly with a tailor, whose customers are in Dagua, El Treinta, and Cali. The police inspector lives outside of El Palmar. The majority of the families in these two groups are in El Palmar but not of it: they are there because of the highway, and form the nucleus of the emerging roadside community. This is not to say that they are unconnected to the community of El Palmar: their presence has very important direct and indirect effects on that community. Nevertheless, their history lies elsewhere than in El Palmar, and a different type of questionnaire would have been necessary to reconstruct that history.

It should be noted that not all households living by the highway were considered to be a part of the roadside community. Many had their roots in the resident agricultural community and were included in the study.

The extreme northwestern corner of El Palmar (see map 6) was excluded from the study because of time and budget considerations. Information on these households was available from the 1973 Population Census (DANE 1973b) and the 1978 Agricultural Census (CENAGRO 1978). From those data it was determined that the exclusion of these families would have little effect on the overall size-distribution of holdings. Although additional cases would have been useful for statistical purposes, this did not justify the additional expense involved in obtaining the information. The response rate among the 106 resident households selected as the community sample was excellent: only one household refused to participate. I also interviewed a sample of nine absentee landowners, excluding emigrants who rarely returned to El Palmar, in order to get a more complete land history for the community and a coverage of the entire spectrum of cultivation techniques and land-tenure arrangements.

## Resident Households Interviewed

There were 207 females and 213 males over nine years of age in the 105 resident households interviewed. The average household contained 5.9 members, generally the nuclear family, although there were some cases of extended family structure. In many households there were brothers or sisters of the heads of household, nieces or nephews, and even unrelated individuals living with the nuclear family. There were cases of grandchildren being raised on the farm while their mothers, usually young and single, worked in Cali (it was not uncommon for girls of fifteen to be sent from El Palmar to the city as domestic workers). For purposes of this study a household, as opposed to a family, was defined as a group of individuals who shared food, since expenditures on foodstuffs constituted the major item in most budgets.

The majority of the households (84 percent) were headed by males. The females heading households were either widows with children, or young, single women with children. The latter were generally receiving minimal support from the children's fathers and had parents living in El Palmar with whom they maintained economic ties. The majority of the young single mothers, however, of whom there were quite a few, lived with their parents and received little or no support from the fathers of their children.

# Appendix C: Agricultural Activity in El Palmar, 1978–79

This appendix supplements the information presented in chapter 6 on agricultural practices in El Palmar during the 1978–79 agricultural cycle. The statistics are drawn from the 105 formal interviews with resident households.

Overall, as table 6.13 of chapter 6 shows, coffee cultivation predominated in El Palmar. For the 105 resident households interviewed 53 percent of all land in use in 1978–79 was under coffee. The rest of the land in use was divided between annual crops (25 percent), pasture (14 percent), and sugar cane and pineapples (8 percent).

## Production for Household Consumption

As chapter 6 demonstrates, agricultural production in 1978–79 was primarily market-oriented. Nevertheless, production for direct household consumption was significant, although none of the households could be classified as "self-sufficient" units. The value of production for home consumption varied from a low of 3 percent to a high of almost 30 percent of total net household income. This variance was attributable more to differences between families in the value of cash income than in the value of retained output. The value of retained output was between 5,000 and 20,000 pesos (40 pesos equaled 1 dollar in 1978–79) for the majority of the households with any land. By way of comparison, a fully employed adult male agricultural laborer could have earned around 30,000 pesos annually in the area of El Palmar in 1978.

The most important component of direct consumption was plantains, which accounted for 2,500 to 10,000 pesos per year for the majority of the households. Plantains were grown as shade trees for the coffee plants and also provided a staple item in the diet. Only the surplus over household needs was sold, despite which plantains ranked as the fourth most important cash crop in El Palmar (see table 6.1). If the value of on-farm consumption is added to the value of marketed output, the total value of plantain production came to around 1,000,000 pesos in the year of the study.

Another staple item was cassava (*yuca*). This crop was grown almost exclusively for direct consumption in El Palmar. Its production was fairly generalized; most commonly it was intercropped on a small scale with other staples such as *arracacha*, maize, and beans. All three of these were grown both for the market and for direct consumption. The value of direct consumption of these four products generally ranged from around 1,000 to 5,000 pesos per household. Overall, approximately an equal proportion of these products was retained for household consumption as was marketed. Their production for direct consumption declined in the 1970s as land was redirected to tomato, green pepper, and coffee production (particularly when *caturra* was planted without intersown food crops).

*Zapallo*, a squash similar to pumpkin, had recently attained considerable popularity in El Palmar due to its ease of cultivation and good flavor. It was being grown on a small scale for household use by many families, along with a scattering of other minor foodstuffs such as *mafafa* (a potato-like tuber). Each household generally had a few fruit trees producing for home consumption, such as banana, orange, guava, lemon, papaya, avocado, and *zapote*.

Those households that grew sugar cane (primarily households with over 7.5 acres of land) always retained some of the *panela* made from it. Whenever tomatoes were grown for the market, a small part of the crop was consumed by the family, although it was difficult to put a value on this consumption. Green peppers were almost exclusively a market crop. Very few families grew carrots, onions, garlic, or herbs such as coriander (cilantro), even though these items were commonly purchased. Carrots were inexpensive and therefore considered not worth growing. Onions were expensive and a basic foodstuff, but required irrigation and fertilizer. The herbs needed to be protected from the chickens and other domestic fowl often found on the farm.

Where there were fowl, these were generally fewer than ten, including chicks, because of the expense involved in feeding them and the high rate of loss (they were rarely enclosed). This was particularly true of laying hens (*gallinas*), which only lay if they receive an adequate diet. For this reason few families could count on a regular supply of on-farm eggs, and even fewer families sold them. At three pesos each, however, eggs constituted a significant product. The chickens themselves were rarely consumed in El Palmar, except at Christmas and sometimes Easter, when chicken is traditionally added to the daily *sancocho* (a kind of soup or stew). A few households had ducks or turkeys, but generally not for home consumption. Minor animals raised for home consumption by a few of the households included rabbits and a type of guinea pig that is larger and meatier than the pet-store variety common in the United States.

Very few households had dairy cows. Pigs were more common, as they could be fed on sugar cane and plantain waste. Both steers and pigs were always sold rather than butchered and consumed on-farm, even though every family bought some meat weekly. No preserving of meat through smoking or drying was carried out in the area.

Almost every household grew coffee, and some coffee was always retained for household consumption. This was generally the damaged beans (called the *guayaba*) that had to be separated out from the good beans so that a higher price could be obtained. Despite ubiquitous coffee cultivation, most families at some time during the year bought roasted and ground coffee in the store, either when the *guayaba* did not suffice or when there was no time to roast and grind it themselves. Many households also occasionally bought plantains. *Panela*, maize, kidney beans, *arracacha*, and cassava were purchased fairly regularly by most households, when not available on the farm. Salt, vegetable lard, and meat were purchased regularly by all households; potatoes, onions, and garlic by most. Other foodstuffs purchased by at least some households included chocolate, refined sugar, carrots, tomatoes, saffron, coriander, lentils, some fruits, eggs, milk (whole or powdered), and pasta.

## Techniques of Cultivation

Tables C.1 and C.2 present data on yields and labor requirements for the major annual crops grown in El Palmar: maize, beans, cassava, *arracacha*, tomatoes, and green peppers. Figures are presented from a variety of sources in order to provide a comparison with the El Palmar data. The last two rows present data for El Palmar: average figures for the community based on the household interview and crop-study data. The crop-study data, available for maize with beans, tomatoes, and green peppers, are based on six case studies carried out in 1978–79. For these studies the participating farmers maintained detailed records of their operations for the specific crop in question.

It can be seen from these tables that there was considerable variation between sources in yields for each crop. Yields for beans varied from 375 to 1,500 kg per hectare, for maize from 875 to 3,500 kg per hectare, for cassava from 4.2 to 25 tons per hectare. The highest figures were

TABLE C.1. *Yields and Labor Requirements of Kidney Beans and Maize*

| | *Kidney Beans* | | | *Maize* | | | *Maize with Kidney Beans* | | | | |
|---|---|---|---|---|---|---|---|---|---|---|---|
| | *Yield (kg)* | *Seed Density (kg)* | *Labor Reqs. (man-days)* | *Yield (kg)* | *Seed Density (kg)* | *Labor Reqs. (man-days)* | *Yield Beans (kg)* | *Seed Density (kg)* | *Yield Maize (kg)* | *Seed Density (kg)* | *Labor Reqs. (man-days)* |
| CVC: *Manual de Costos* 1977b | 1,400 | 80 | — | 3,500 | 20 | — | — | — | — | — | — |
| DAP: *Anuario Estadístico del Valle del Cauca* 1978 (1976 data) | 1,129 | — | — | 2,900 | — | — | — | — | — | — | — |
| FEDECAFÉ: *Manual de Instructivos* (for Caldas) 1963 | 900–1,500 | 60 | 68 | 3,500 | 33 | 75 | — | — | — | — | — |
| FEDECAFÉ: *Campañas Técnicas* 1975, "technified" | 1,000 | — | — | 2,600 | — | — | — | — | — | — | — |
| FEDECAFÉ: *Campañas Técnicas* 1975, "traditional" | 400 | — | — | 1,200 | — | — | — | — | — | — | — |

| | | | | | | | | | | | |
|---|---|---|---|---|---|---|---|---|---|---|---|
| PRODESARROLLO: *Zona Cafetera Central* 1972b (1969 data) | 660 | — | — | 1,270 | — | — | — | — | — | — | — |
| MINAGRICULTURA: *Cifras del Sector Agropecuario* 1979 (1977 data) | 676 | — | — | 1,365 | — | — | — | — | — | — | — |
| INCORA: *Ingresos y Costos en Zonas de Crédito* 1966 | 375[a] | — | 39[a] | 1,700 | — | 50 | 400 | — | 2,250 | — | 83 |
| Nariño 1979 (1975–76 data) | 616 | — | — | 1,100 | — | — | — | — | — | — | — |
| ICA: García Rovira 1973 | 445 | — | 31 | 875 | — | 37 | — | — | — | — | 62 |
| El Palmar, average, 1978–79 | 470[b] | — | 50 | 1,250[b] | — | 40 | 470[b] | — | 1,250[b] | — | — |
| El Palmar, crop studies, 1979 | | | | | | | | | | | |
| Case 1 | — | — | — | — | — | — | 250 | 13 | 1,500[c] | 17 | 70 |
| Case 2 | — | — | — | — | — | — | 460 | 50 | 2,150[c] | 17[d] | 65 |

*Note:* All figures are per hectare.
[a] In mountain areas only, all manual operations.
[b] Average, planted separately and intercropped (maize with beans).
[c] *Choclo:* still moist; sold on the ear.
[d] Improved seed.

TABLE C.2. *Yields and Labor Requirements of Other Annual Crops Grown in El Palmar*

| | *Cassava* | | | *Arracacha* | | | *Tomatoes* | | | *Green Pepper* | | |
|---|---|---|---|---|---|---|---|---|---|---|---|---|
| | *Yield* | *Seed Dens. (gm)* | *Labor Reqs. (man-days)* | *Yield* | *Seed Dens. (gm)* | *Labor Reqs. (man-days)* | *Yield*[a] | *Seed Dens. (gm)* | *Labor Reqs. (man-days)* | *Yield* | *Seed Dens. (gm)* | *Labor Reqs. (man-days)* |
| CVC: *Manual de Costos* 1977b | 11 tons | 10,000 | 118 | — | — | — | 27.5 tons (1,750 crates) | 450 | 435 | — | — | — |
| DAP: *Anuario Estadístico del Valle del Cauca* 1978 (1976 data) | 8,200 kg | — | — | — | — | — | 23.5 tons (1,500 crates) | — | — | — | — | — |
| *FEDECAFÉ: Manual de Instructivos* (for Caldas) 1963 | 15–25 tons | 7,000 | 106 | — | — | — | 35 tons (2,225 crates) | 350 | 256 | — | — | — |
| FEDECAFÉ: *Campañas Técnicas* 1975, "technified" | 20,000 kg | — | 100 | — | — | — | — | — | — | — | — | — |

| | | | | | | | | | | | | |
|---|---|---|---|---|---|---|---|---|---|---|---|---|
| FEDECAFÉ: *Campañas Técnicas* 1975, "traditional" | 9,000 kg | — | 100 | — | — | — | — | — | — | — | — | — |
| PRODESARROLLO: *Zona Cafetera Central* 1972b (1969 data) | 6,680 kg | — | 115 | — | — | — | — | — | — | — | — | — |
| INCORA: *Ingresos y Costos en Zonas de Crédito* 1966 | 8,125 kg | — | 78 | 10,937 kg | — | 217 | 1,200 crates (19 tons) | — | 334 | — | — | — |
| Nariño 1979 (1975–76 data) | 11,000 kg | — | — | — | — | — | — | — | — | — | — | — |
| El Palmar, average, 1978–79 | 4,200 kg | — | 90 | 6,250 kg | — | 125 | 425 crates (6.7 tons) | — | — | 260 crates | — | — |
| El Palmar, crop studies, 1979 | | | | | | | | | | | | |
| Case 1 | — | — | — | — | — | — | 2,500 crates | 625 | 790 | 310 crates | 415 | — |
| Case 2 | — | — | — | — | — | — | 1,800 crates | 833 | 830 | 265 crates | 415 | 220 |

*Note:* All figures are per hectare.
[a]Conversions calculated using approximate weight per crate of 16 kg; 1 long ton = 1,016 kg.

presented in the first four studies. These figures were yields for "commercial" production (i.e., using the latest technology and flat, fertile soil). They were anticipated yields based on agronomic research and controlled trials, rather than actual yields reported from the field. The yields were considerably higher than those reported in the remaining studies, which tended to be based on actual field situations or were estimates based on actual, rather than ideal, techniques and terrain corresponding to the situation in the mountain zones.

In the following discussion of techniques of cultivation for specific crops, reference is frequently made to the *plaza*, the local unit of land measure. One *plaza* equals 6,400 square meters (.64 hectare, or approximately 1.5 acres).

## *Kidney Beans*

The average yield in El Palmar was above the median yield in the INCORA study, the average yield in the García Rovira study, and the "traditional" yield in the FEDECAFE study in table C.1. It was below the yields given in the other studies.

There was only slight variation in the way that kidney beans were grown in El Palmar. Beans were generally planted with maize rather than alone. The seed was "unimproved," sometimes from the previous crop and sometimes purchased. The practice in the area was to plant two seeds per hole. The major variation in techniques of cultivation was in the type and quantity of fertilizer and pesticides used. Most farmers interviewed purchased *gallinaza*, an organic fertilizer made from chicken droppings, for their beans. A few used chemical fertilizer. Those who grew tomatoes or green peppers often followed those crops with beans, so that the beans would get the benefit of the heavy applications of chemical fertilizer used for the earlier crops. This was true of case 2 in the maize-bean crop study, while case 1 used only *gallinaza* and the land had not been previously fertilized with chemical fertilizer. Only recently have the farmers of El Palmar begun to apply any type of fertilizer to their beans, although the practice is now widespread. It "wasn't necessary" before, but the soil has apparently deteriorated. Only a few farmers sprayed their beans with pesticides, and the type and quantity varied considerably.

Labor requirements were very low for kidney beans, particularly if they were grown with maize for support as they climbed, and if they were not sprayed. Labor time could be as low as 21 man-days per *plaza* (35 per hectare) if the land did not have to be cleared first, 41 man-days per *plaza* (68 per hectare) if it did. The beans were ready for harvest in approximately two and a half months. Labor requirements were concentrated in the sowing (4–5 man-days per *plaza*), one weeding (5–10 man-days) approximately three weeks after planting, and the harvest (12 man-days). If labor was not available to harvest the beans when the pods were still colored (*pintado*), in which case they were sold in the pod, the beans had to be left to dry on the vine and later threshed, adding 3 to 4 man-days to the labor requirements. The timing of the tasks for kidney bean cultivation in El Palmar is presented in Reinhardt (1981 : 388, table 8.4–12), along with the cultivation schedule for maize, tomatoes, green peppers, and coffee.

## *Maize*

Average maize yields in El Palmar were close to or above the yields reported in the comparable studies (5 through 9). Maize was generally grown with beans, although it was also planted by itself in small garden plots, or scattered with other food crops (cassava, *arracacha*, plantains, *mafafa*, etc.).

There was little variation in the way maize was grown. It was rarely fertilized or sprayed (although if grown with beans it was exposed to the fertilizer and the pesticides applied to the beans). The seed was generally "unimproved," although in a few cases "improved" seed was used. The practice, as with beans, was to plant two seeds per hole.

The labor requirements for maize in El Palmar were even lower than for beans: as few as 12 man-days per *plaza* (20 per hectare) if the land did not have to be cleared, 32 man-days per *plaza* (53 per hectare) if it did. The only operations were the planting (4–5 man-days per *plaza*) and the harvest (8–10 man-days). The maize was ready to be harvested as *choclo* (moist) in approximately four and a half months. If labor was not available during the one to two weeks that the maize could be picked as *choclo*, it had to be left to dry on the stalk for another one and a half to two months.

## *Cassava*

Cassava yields in El Palmar were very low. As this was almost exclusively a food crop, purchased inputs such as fertilizer or pesticides were rarely used. Cassava yields had fallen considerably over the years, according to the farmers, because it was "hard on the soil" (cassava requires many nutrients and rapidly depletes the soil). Because good yields could only be obtained on land that had lain fallow or with the use of purchased fertilizer, cassava cultivation had declined in El Palmar.

Labor requirements for cassava were higher in El Palmar than for maize or beans: approximately 55 man-days per *plaza* (90 per hectare), including clearing the land. Labor requirements were heavily concentrated in the soil preparation and planting period: approximately 20 man-days per *plaza* to clear the land, and 20 to dig the holes and plant the cuttings. During the fourteen- to sixteen-month growing period, the only operations were two weedings of around 5 man-days each per *plaza*, and finally the harvest (around 10 man-days per *plaza*). Cassava was usually grown on a small scale as a home crop, either by itself or frequently inter-planted with *arracacha* and sometimes maize and beans. It was harvested as needed by the household, generally by the women and children (although cassava rots quickly after harvesting, it remains good in the soil for a prolonged period).

## Arracacha

As with cassava, *arracacha* yields in El Palmar were quite low without the use of fertilizer. Unlike cassava, however, *arracacha* was often grown as a cash crop (see table 6.1), in which case fertilizer was applied, generally *gallinaza*.

Labor requirements for *arracacha* were slightly higher than for cassava because of the application of fertilizer (during planting) and because of two more weedings. The timing of operations, except for the extra weedings, was similar to that for cassava.

## *Tomatoes*

Tomato yields in El Palmar were low on average, but this average masked a wide variation: from cases where the entire crop was lost to pests, to extraordinary yields of 1,800 and 2,500 crates per hectare in the two case studies (twenty-eight to forty tons, approximately). These high yields were perhaps due to the high seed density, and to the small plot sizes (approximately .25 *plaza* each) and continuous and careful attention to the crop, that commonly characterized tomato cultivation in El Palmar.

There was copious application of fertilizer, both organic (*gallinaza*) and chemical, and of pesticides, with little variation between farms. Outlays for pesticides and fertilizer alone for a quarter *plaza* in tomatoes amounted to some 6,000 pesos or more (24,000 pesos per *plaza*, or 80 percent of the annual salary of a fully employed adult male agricultural laborer, although plots of this size were rare). There were also outlays for seed, and for bamboo for stakes and crates if not available on the farm. A sprayer for the application of the pesticides and a hose for irrigation were also needed. Despite the outlays on pesticides, there was considerable crop loss to pests in many cases. These losses explain the low average yield.

Tomatoes were the most labor-intensive crop grown in El Palmar. They required almost constant attention from the time the seedlings were planted to the harvest: irrigation (done by hand with a hose: sprinklers would damage the leaves), staking, tying the plants to the stakes as they grew, weeding, and spraying. The crop was sprayed two or three times a week until the beginning of the harvest, which started approximately two and a half months after the seedlings were planted. The harvest itself lasted for six to eight weeks, during which time the plants were sprayed once or twice a week, tying continued, and two passes were made weekly through the crop to pick the ripe tomatoes. Altogether the tomato crop required some 480 man-days of labor per *plaza* over a period of four and a half months, which would occupy five men almost full-time. This high labor requirement, plus the expenses involved in purchasing the necessary inputs and the riskiness of the crop, explain the generally small size of the tomato plots in El Palmar.

## *Green Peppers*

None of the sources consulted contained information on the cultivation of green peppers in Colombia. It is therefore not possible to compare yields in El Palmar to those obtained elsewhere, except to say that they appeared to be quite low. In both of the case studies, the plants became diseased midway through the study, and yields were obviously reduced considerably. Nevertheless they were higher than the average for El Palmar in 1978–79, indicating that crop loss to pests was extensive throughout the community for this crop. Because of this problem, which has apparently been worsening, cultivation of green peppers has been declining in recent years in El Palmar. Not only was it apparently riskier than tomatoes, while requiring only slightly lower expenditures on fertilizer and pesticides, but the price was lower. In the case studies, the green peppers sold for 80–100 pesos per crate, while the tomatoes sold for 100–240 pesos per crate (note the wide fluctuations in price for these crops over the six-week harvest period).

The differences in price may have been at least in part because of differences in labor requirements. Labor requirements for green peppers were considerably below those for tomatoes in the one case for which information was available, although considerably above those for the other annual crops. The figure for green peppers understates the normal labor requirements of this crop, however: once the plants became diseased and failed to respond to treatment, fumigating and weeding were stopped. A higher yield would have also required more labor for the harvest. If the plants had not become diseased, the final figure for labor requirements would have been higher, although still less than for tomatoes because staking and tying are not required for the green pepper plant. As with tomatoes, green peppers were normally irrigated, weeded, and sprayed with pesticides regularly until, and even during, the harvest. The harvest began approximately two months after the seedlings had been planted and lasted for six to eight weeks.

## *Coffee*

The yields and labor requirements for coffee and other perennial crops varied depending on the stage in the life cycle of the plants. In addition, for coffee there was considerable variation between farms in El Palmar in the techniques of cultivation, with concomitant variation in labor requirements and yields.

The traditional method of coffee planting consisted of digging a small hole or pulling up an old coffee plant, and dropping a berry in the hole. This method was still used in 1978–79 by many farmers in the area. The modern method involved germinating seedlings, transplanting them into black plastic bags filled with dirt, and then later planting the small coffee plants in the coffee plantation. The timing of the planting was important: if the farmer waited too long, the roots passed through the plastic bags and into the ground, after which time it was not possible to move the plants without damaging their root systems. One farmer in El Palmar was observed to lose 2,000 seedlings because he was not able to obtain labor for the planting during the necessary two-week interval (which fell during the peak of the coffee harvest).

Both the traditional *arábica* varieties and *caturra* were being planted in El Palmar. The traditional varieties give their first harvest after three years. They are long-lived plants (some in El Palmar were still producing, although with quite low yields, after fifty years or more). *Caturra* gives its first harvest more quickly (after one and a half years), and with the necessary complementary inputs it yields more abundantly, but its lifespan is much shorter. It also should be pruned back after approximately eight years, resulting in a one-year loss of output.

Although *caturra* supposedly requires less shade—in fact, it is expected to reach its full yield potential only with little or no shade—the farmers of El Palmar tended to use the same shading pattern with both varieties. The leguminous *guamo* tree was used as permanent shade. This is a tall, broadleaf tree, which was planted at wide intervals around the plantation. The plantain tree, lower and more densely planted (100 per *plaza* of coffee), formed a second, temporary layer of protection. The wisdom of using this system of shading with *caturra* was demonstrated during the course of this study. One of the seasonal dry periods was quite severe. The sun can be quite intense at an elevation of 5,000 feet near the equator. Some stands of new *caturra* trees that had been planted with little or no shade weakened and became diseased during the dry spell, while stands of new *caturra* trees that had been planted with the traditional shading pattern were unaffected.

The *guamo*, in addition to providing shade, fixes nitrogen into the soil. Traditional coffee cultivation also involved the use of organic fertilizer in the form of compost from the plantain and *guamo* trees and the coffee pulp that had been removed from the coffee beans during processing and left to decompose and ferment for several months. All coffee farmers in the study used these forms of organic fertilizer. In addition, many coffee growers were using *gallinaza* and/or chemical fertilizer at the time of the study (including all of the farmers growing *caturra*). There was wide variation in the frequency of fertilizer applications and in the quantity applied.

In addition to differences in the variety of coffee grown, the method of planting, the type and quantity of fertilizer applied, and the amount of shading, there were also differences between farms in the density of the coffee plants, in the use of pesticides, in the frequency of weeding, and in the pruning of the coffee trees. Labor requirements per *plaza* varied considerably because of these differences in techniques of cultivation and (related) differences in yield.

Harvest labor accounted for the bulk of the labor requirements for coffee cultivation, even with the additional labor requirements of the "modern" method of cultivation. As with tomatoes and green peppers, the coffee harvest involved repeated passes through the plants, since the berries ripen over the course of several months and must be picked within a few days of full ripening. On one farm for which a detailed study was carried out, a total of 141 man-days of labor were applied to one *plaza* of five year-old *caturra* (2,000 trees). Of this, 116 man-days were for the harvest, while the remaining 25 were for one weeding and four applications of fertilizer. In addition, approximately two hours were required per *arroba* of coffee for processing (depulping, washing, sorting, and drying). In the above case, with 115 *arrobas* of coffee harvested, this came to another 29 man-days of labor, for a total of 170 man-days per *plaza* (270 per hectare).

This compares to the figures calculated by FEDESARROLLO (1978) of 98 man-days per hectare under production for the "traditional" method, and 315 man-days per hectare under production for the "modern" method. The yields given by FEDESARROLLO were 24 to 31 *arrobas* per *plaza* for the traditional *arábica* variety, and 96 to 120 *arrobas* per *plaza* for *caturra*. In El Palmar average yields per farm for the resident-owned farms ranged from 2.3 to 101 *arrobas* per *plaza* of coffee.

### *Sugar Cane*

Sugar cane was cultivated in El Palmar for processing into unrefined block sugar (*panela*). Detailed information was not obtained on labor requirements and yields for sugar-cane cultivation and *panela* production. The labor requirements were concentrated in the harvesting and processing operations under the traditional system of cultivation, with little labor expended during the maturation period of the plant. Male workers cut the ripe cane by machete. Wood was gathered to begin the cooking of the cane juice; once the process was under way, the cane stalks themselves, with the juice extracted, became the fuel.

The processing of sugar cane into *panela* required a press to extract the juice from the cane and large copper kettles to boil it down. The presses used in El Palmar consisted of two meshed rotating discs, which were turned by a horse or mule while the cane was manually fed (often by a child, although the work is dangerous) between them. The juice was caught below the presses and ran through bamboo conduits to the series of kettles in which it was then cooked. Because of the equipment and the amount of cane needed to produce *panela*, it was not profitable on a small scale. This explains the concentration of *panela* production on the larger farms in El Palmar. Some of the small farms did, however, grow cane on a small scale for home use and pay a neighbor, usually in shares, to process the cane into *panela*.

The pressing itself was not begun until all the cane to be processed had been cut. Once started, processing was generally continued day and night until all the cane was pressed and all the syrup cooked, some two to four days depending on the amount of cane. On farms with over a *plaza* of mature cane, processing would be carried out in batches once every two weeks until all the ripe cane had been harvested.

## Division of Labor

The predominant patriarchal structure was reflected in a well-defined division of labor along age and gender lines in El Palmar, although exceptions to the general rule did exist. Household chores—cooking, bringing meals to the field if necessary, cleaning, washing and ironing of clothes, child care—were the responsibility of the women. In many of these chores, they were aided by the young children of both sexes, but particularly the girls. By the age of nine or ten, young boys began to be relieved from assisting in these chores, while the responsibility of young girls began to increase. There were cases of girls of eleven or twelve having full responsibility for household food preparation. By their early teens, the boys became more fully involved in the "male" agricultural activities.

In addition to household chores, women and children in farm households performed a variety of other tasks. They maintained and harvested garden plots and small plots of household food crops. They raised fowl and rabbits for household consumption or sale. They sometimes cared for pigs. They participated in the harvest of all traditional crops except sugar cane. They were primarily responsible for the processing of the harvested coffee beans (sometimes depulping, washing and sorting, and drying). Women's involvement in the newer cash crops of tomatoes and green peppers was far less than their involvement in the traditional crops. In many cases they did not even participate in the harvest of the new crops.

The men and teenaged boys in agricultural households in El Palmar generally carried out the soil preparation for all crops, the planting of the cash crops, the fertilizing, fumigating, staking and pruning, the cutting of the bunches (*racimos*) of plantains for household consumption or for sale, the harvesting of sugar cane and the preparation of *panela* from it, and the tending of the larger animals (pigs, cows, horses). They carried out the harvesting of other crops along with the women and children, as well as some of the processing of the coffee beans.

The male head of the household brought the cash crops to market and either purchased the household provisions himself or made the allocations for those provisions. He made all decisions about agricultural activities and the disposition of household income. Even land inherited by the wife was frequently put in the husband's name and worked by him. The women traditionally have had little access to the cash income of the household.

## Marketing of Agricultural Products

Coffee was almost always brought to Dagua by the farmers on the "popular" buses. In Dagua the majority of the farmers from El Palmar sold their coffee to the Coffee Cooperative (the marketing arm of the National Coffee Federation). A few were selling to other merchants at the time of the study, a couple because the merchant had advanced them credit during the previous year, the rest because they were angry about the Federation's TAC program. Under this inflation-fighting program the cooperatives had been issuing certificates, redeemable after six months, as part payment for each sack of coffee they purchased. At the time of the study, the cash payment per *arroba* (12.5 kg sack) of coffee was 650 pesos, and the certificate was worth another 50 pesos per *arroba*. The independent coffee merchants were paying approximately the same cash price, without the additional certificates. Nevertheless, some farmers were selling to the independents because they simply did not want to deal with the certificates or felt that they were being tricked by the cooperative. Many of the farmers selling to the cooperative did not really understand the value of the certificates and the redemption process, and had lost them, thrown them away, or kept them past the redemption period.

Plantains were marketed by the roadside in El Palmar, secondarily in El Treinta and Dagua, and in a few cases in Cali. They were sometimes sold to the consumers by the farmers themselves and sometimes "on consignment" to the marketer. The same was true of many other subsistence crops: maize, *arracacha*, cassava, *panela*, fruits. Kidney beans were usually sold to the general stores in El Palmar and in Dagua, or to a middleman in El Palmar. Eggs were generally sold directly within the community or to the local stores. Milk was also generally sold within the community, picked up each morning by the consumers. A few members of the community were purchasing milk and eggs from the local producers for resale within the community. Livestock was sold to the butcher stalls by the roadside and in El Treinta. Tomatoes and green peppers were sold through intermediaries to the markets in Cali.

# Appendix D: Calculation of Modernization Index

TABLE D.1. *Calculation of Modernization Index*

| *Indicator* | *Scores Assigned* | *Maximum Score* |
|---|---|---|
| Coffee depulper, mechanical | 15 | 15 |
| Coffee dryer (hot-air) | 20 | 20 |
| Generator | 10 | 10 |
| Pesticide applicator (manual sprayer) | 10 | 10 |
| Water sprinkler | 5 | 5 |
| Wagon | 5 | 5 |
| Manual chopper for fodder | 5 | 5 |
| Water tank | 5 | 5 |
| Vehicle (car or truck) | 15 | 15 |
| Store or shop | 10 | 10 |
| Enclosed chicken yard | 5 | 5 |
| Beekeeping | 10 | 10 |
| Six or more pigs | 10 | 10 |
| Use of bank credit (past or present) | 15 | 15 |
| Use of credit for farm improvement | 5 | 5 |
| Current indebtedness: | | |
| 25,000–50,000 pesos | 5 | |
| 50,000–100,000 pesos | 10 | |
| 100,000 or more pesos | 15 | 15 |
| Loaned money for interest | 10 | 10 |
| Planted *caturra* | 10 | 10 |
| Consultation with Coffee Federation agents | 5 | 5 |

TABLE D.1. *(continued)*

| *Indicator* | *Scores Assigned* | *Maximum Score* |
|---|---|---|
| Use of purchased fertilizer for coffee: | | |
| Currently use | 10 | |
| Use partly, or plan to start next year | 5 | 10 |
| Purchased improved seeds | 10 | 10 |
| Purchased agricultural chemicals | 5 | 5 |
| Cash outlays per plaza: | | |
| 5,000–10,000 pesos | 15 | |
| 10,000 or more pesos | 20 | 20 |
| Maximum obtainable score | | 230 |

# Appendix E: Case Studies

The cases presented here are based on the formal household interviews from 1978–79 and on oral histories obtained in 1978–79 and 1983–84. The first section discusses the economic status of seven households up to 1978–79. The second section looks at changes in the economic status of nine households between 1978–79 and 1983–84.

## Economic Status to 1978–79

### *Case 1: The Coffee Farmer*

In 1955, when Arcesio Benavides married Diomelina Álvarez and received 3 acres of land from his father Aristides, the land had 1.5 acres in coffee. Arcesio planted more coffee, using beans from the existing trees and simply dropping them in a small hole with no fertilizer or other preparation. In 1958 this family lost half its land to the construction of the highway, but Arcesio talked his father into selling him another 4.5 acres. Part of that land was in sugar cane, which he kept, using his father's *trapiche* to make *panela*. The rest of the land he began to plant in coffee and basic foodstuffs.

In the 1960s, with 6 acres and a growing young family, Diomelina and Arcesio struggled to make ends meet. Much of the sugar cane and coffee was old and yields were low. The family spent most of its income on food and the rest on other necessities. At one point Diomelina borrowed some money from a relative, which she invested in part-ownership of a small *cantina*. Her customers bought on credit, however, and the business folded. By 1970 the oldest son was working full-time on the farm, despite Diomelina's desire for her children to finish their education. The next son, aged fourteen, was studying in Dagua and helping on the farm when he was home. The twelve-year-old daughter was in school in El Palmar, but helped her mother with household chores, including the care of her five-year-old cousin and one-year-old brother. She and her mother helped in the harvest and processing of the coffee and the food crops.

Part of their farm was too steep to be cultivated. The rest was intensively farmed in coffee and sugar cane, with scattered plots of food crops. The family used no commercial credit, no hired labor, and no commercial inputs. But changes were beginning to take place in the community. The Coffee Federation was recommending that coffee farmers plant *caturra*. The Federation would provide technical assistance, and credit through the Banco Cafetero, to small farmers interested in planting *caturra* using their recommended methods. Diomelina was interested, but Arcesio was unwilling to take some of their scarce land out of production for a year and a half. (This illustrates the difficulty, discussed in chapter 6, that farmers in the 3.7–7.4 acre category faced in taking advantage of the new opportunities of the 1970s.)

In 1973 Diomelina received her inheritance from her father, Juan: 3.5 acres of land containing the long-unoccupied Álvarez family house, in need of considerable repair, and a small plot of equally neglected coffee trees. The rest of the land was uncultivated and overgrown with dense tropical shrubbery. Diomelina wanted to plant *caturra* on her land. However, she did not yet have her land title (this was a lengthy and costly process), and was not eligible for a bank loan without it. Arcesio agreed to use his land title, and they obtained a loan in 1973 from the Banco Cafetero. Relying totally on family labor they cleared 1.5 acres of Diomelina's land and planted 2,000 coffee trees, principally *caturra*. When the new trees began to produce eighteen months later (more quickly than the three years needed for traditional varieties), they caught the upsurge in coffee prices resulting from Brazilian frosts in the mid 1970s.

Family income rose. Diomelina obtained formal title to her land, despite pressure from Arcesio to title the land under his name. In late 1977 Diomelina secured another loan to further develop her land. With the loan funds and earnings from the new coffee in production, they planted another 1,200 trees. This planting was much more costly than the first, both because fertilizer prices had risen considerably in the intervening four years and because they were now having to rely increasingly on hired labor. The second son had gone to work in Dagua, while the increased scale of their farm demanded a greater labor input.

Arcesio's older coffee plantation was also upgraded with careful attention, scattered replanting, and the use of fertilizer. In the 1978 coffee harvest, the output of the combined operations doubled over the 1977 level. Production costs also increased. In the 1978–79 agricultural year, they spent 50,000 pesos (U.S. $1,250 at the 1978–79 exchange rate of 40 pesos to the dollar) on fertilizer, chemicals, and hired labor, principally for the coffee plantation. Arcesio continued to cultivate sugar cane for *panela*, but dropped all foodstuff production because of the shortage of family labor. Their net income from farm operations in 1978–79 was substantial: 178,000 pesos, plus 7,000 pesos worth of household consumption of farm products. Cash income was primarily from coffee and plantain sales, as *panela* prices fell so low in early 1979 that they began to grind cane for home consumption only. Coffee earnings promised to increase the following year as output continued to grow—although expenses, too, would continue to rise.

In 1978, on the strength of their increasing income, they took out a 100,000-peso loan to build a combined living quarters/coffee drying unit, consisting of two brick-and-cement bedrooms with a large flat cement roof on which coffee could be dried. Attached to this structure were two tile roofs on rollers that could be quickly pushed over the coffee if it started to rain, or at night. This improvement was necessary to handle the increased volume of coffee the farm was producing. In addition, they built a *beneficiadero* with the loan funds and with coffee earnings. This coffee-processing unit was devised by putting the hand-operated depulper into a concrete base containing tanks to wash the depulped beans. The family also acquired a horse and miscellaneous farm implements, such as a wheelbarrow and hoses.

The family's consumption standards improved markedly. With the loan funds, a roomy kitchen was built next to the new bedrooms, as well as an outdoor cement wash-station, shower, and toilet. In 1977 they obtained electricity, brought into the area by the CVC following the construction of the highway. The son working in Dagua bought the family a large TV set. Their diet was ample and nutritious: the family was spending above its basic subsistence level on food purchases. With Diomelina's insistence, the family continued their outlays on education. The oldest daughter went to school in Cali, and then stayed there to work for the telephone company. The two youngest children were in school, the boy in El Palmar, the girl going to the high school in Dagua.

In sum, the family's increasing income was spent on improved diet, home and farm improvements, consumption items, education, and miscellaneous expenses such as medical care. Part of these outlays had been financed by loans. The increased production of the farm in 1978 generated enough cash earnings above their basic requirements to have paid off all outstanding loans. Instead, the family increased its outlays on food and other items, increased its indebtedness even further, and continued with home and farm improvements.

In addition to adopting *caturra*, members of the family responded to other suggestions made by the Coffee Federation and agents from several government programs. The oldest son took out a loan to begin a beekeeping operation on a small plot of his father's land. Diomelina obtained a loan to construct a pigsty and began to raise pigs. She hoped to earn more money to continue to cover the education expenses of her youngest children. The family did not, however, extend their land holdings in this period: according to Arcesio, "one does not need much land, only that it be well-tended." Land prices in El Palmar were increasing rapidly with the expanding agricultural activity of the 1970s.

The story of this family has been told in some detail because it illustrates several important features of the transformation of coffee cultivation in El Palmar. This transformation increased labor demand for coffee cultivation and especially for the harvest, both because of the increased amount of land in coffee and because of the increased labor requirements per acre in coffee. Second, new inputs, especially fertilizer, but also equipment, were necessary. Third, commercial credit was important to the new operations, especially because of the increased labor requirements and the need for purchased inputs and equipment. And fourth, it was difficult for full-time farmers with small holdings to adopt the new coffee variety, because it would mean taking some of their land out of operation for eighteen months.

### *Case 2: The Tomato Trader*

Desiderio Almeida was born in 1943 in La Virgen, where his parents had a small farm. When he was in his teens, in the early 1960s, he left home and went to Quindío to look for work on the coffee farms. In 1964 he met Rubiela, who was from Caldas and was working on the same coffee hacienda. In 1966 they briefly returned to La Virgen, where Desiderio worked *en companía* for an uncle, but they were unable to afford even a lot for a house and soon went back to Quindío. There they saved some money, and in 1968 returned to La Virgen, where they bought a small plot of land (1.5 acres).

Desiderio worked as a *jornalero*, and after a year they built a house. By this time the new highway was open and farmers in the area were beginning to plant new crops for the Cali market. Desiderio planted tomatoes on his land and sold the first crop to a middleman. The Cali market structure was very different from the produce markets in El Treinta or Dagua, where the farmers dealt with small vendors who were frequently friends or relatives, or where the farmers sometimes marketed their produce directly to the consumer. In Cali the produce could be sold to the supermarkets opening up in the wealthier sections of the city, to the large mass markets (*galerías*), or to the large wholesale markets out by the airport. These were highly competitive markets and difficult to break into. Prices fluctuated continuously, and buyers often rejected produce either because they were overstocked or because they judged it to be substandard. The next harvest, Desiderio went to Cali with a friend who knew how to operate in the urban markets. From then on Desiderio marketed his tomatoes himself.

In 1971 they sold the house and lot in La Virgen for 11,000 pesos and moved down the mountain to El Palmar, where Desiderio bought 2.5 acres of land from his godfather (*compadre*). He paid 30,000 pesos for the more valuable land in El Palmar—15,000 pesos down and 2,500 pesos a year for the remainder. Desiderio continued to plant tomatoes and also began to buy produce from his neighbors to resell in Cali, bringing the produce to the city in the transport buses that serviced the communities along the highway. He worked *en companía* with some of these farmers, advancing them fertilizer and pesticides, which he bought in Cali, and then deducting the cost from his payment for their produce.

Tomato cultivation and his trading business were profitable. In 1976 Desiderio bought an 18-acre farm and another of 4.5 acres, financed with his savings and with loans. In 1978 he bought a 6-acre farm, giving him a total of 31 acres—a sizeable holding for El Palmar in this period. By this time Desiderio and Rubiela had four children, but the oldest was only twelve. Desiderio had a fairly diversified farm, with sales of coffee, plantains, maize, beans, pigs, and tomatoes. He was dependent on hired labor: *jornaleros* for the coffee and other crops, and one permanent worker who farmed some of Desiderio's land *en companía*. Desiderio had farm expenses in 1978–79 of 70,000 pesos for seed, chemicals, feed for the pigs, and hired labor. His farm netted him 160,000 pesos, including the *en companía* earnings and the value of farm produce consumed by the family. His business netted him approximately 150,000 pesos that year, for a total of over 300,000 pesos—far more than the 60,000 pesos they spent that year on food for the household and the workers.

Desiderio had spread himself thin, however, with his land purchases. He still owed money on the farms. In addition, he had just taken out an 80,000-peso loan with the Banco Cafetero to plant new coffee on his land. He wanted to buy a truck for his business, but did not have the cash. His earnings were invested in the land and in his business, supplying his producers with their inputs. He had come a long way, however, from the landless migrant worker of the 1960s. Only thirty-six, he had a sizeable farm and a prosperous business, both of which he hoped to improve. He was definitely a young man on the rise.

As the case of Desiderio Almeida indicates, the economic transformations taking place in El Palmar went beyond the adoption of new crops and technology. Although Desiderio had gotten his start by cultivating tomatoes, he soon had turned to the even more lucrative venture of marketing the new crops in Cali. His story illustrates in a dramatic fashion the types of opportunities opening up in the 1970s not only for landed households (and we can count Arcesio and Diomelina, meager as their holdings were, in this category), but also for land-poor households. Two other cases illustrate the broader ramifications of the 1970s transformation for these households.

### *Case 3: The* Mayordomo

We have looked at two paths to success in the El Palmar of the 1970s. A third was, surprisingly, certain forms of agricultural labor. The coffee boom of the 1970s attracted new migrants to El Palmar and also absentee landowners, generally from the cities of the valley, looking to acquire small to medium coffee farms. These absentee coffee growers needed managers to run their farms—called *agregados* on the smaller or more traditional farms and *mayordomos* on the larger or more modern ones.

Omar Gomez was an agricultural laborer who came to El Palmar in 1973, at the age of twenty-nine, in search of work. He found employment as a *jornalero*. Within a few years he had been offered the job of *mayordomo* on a 28-acre coffee farm recently purchased by a Cali resident. This absentee owner's main business was wholesaling at the Calabasa market, and he soon opened up a small store on his farm, where he sold rice, coffee, soft drinks, chocolate, and a variety of other items to the residents of El Palmar. Omar was given the job of managing the store as well as the farm. In addition to his salary of 36,000 pesos a year, he received a commission on store sales. Omar married Marta Garcez, and she was given the job of feeding the farmworkers. They were paid a fixed amount per worker and were able to make money on the operation.

In 1978–79 Omar and Marta earned 116,000 pesos, net of all food expenses. This was a considerable amount for a childless young couple, and their savings were rapidly growing. Omar hoped soon to be able to buy land of his own in El Palmar.

### *Case 4: The* Cantina *Owner*

Another young man who got his start as a *mayordomo* was Norberto Bermúdez. An urban worker, he had moved to the country in 1970 in response to an offer to manage a farm in El Palmar. He brought his wife and two young children with him. Their savings grew quickly, and Norberto acquired a small plot of land by the highway, where he decided to open a business. By 1978 Norberto had a busy *cantina*/poolhall, with an outdoor kiosk for dancing, and the family lived on the second floor of the new brick building that housed his business. He had acquired another house by the highway, which he rented out, and he had become a moneylender in the community. His son and daughter attended high school in Dagua. The family's income from the *cantina* and house rental (no information was available on the income from moneylending) was 100,000 pesos in 1978–79, more than sufficient for their needs.

### *Case 5: The Sharecroppers*

Two other young men on the rise were Favio and Bolívar Cabrera. After their brief migration to Quindío in 1965, they had returned and worked with Lubino on his land in the late 1960s and early 1970s. In 1973 they rented 1.5 acres of land from an elderly resident of El Palmar, planting traditional foodstuffs (*arracacha*, maize, and beans). With the earnings they bought two calves (7,000 pesos for the two), which they fattened on rented pasture and sold. The next year they rented 1.5 acres of land in El Palmar from the large traditional absentee landowner discussed in chapter 6. He liked the way they worked, and offered them land *en compañía*—he would provide the land, they would provide the labor, and all would share other expenses (seed, fertilizer, etc.). Within a few years, Favio and Bolívar were farming 15 acres in basic foodstuffs *en compañía*. The maize and beans they sold in local markets; the *arracacha* was sold by their brother Carlos in Cali.

With the earnings from sharecropping and the sale of the cows, Favio and Bolívar were able in 1977 to buy 75 acres of pasture land in El Queremal. They paid 75,000 pesos, or 1,000 pesos per acre—considerably more than the 35 pesos per acre that Aristides had paid for land in El Queremal in the early 1930s, but still far less than the value of land in El Palmar. It is interesting to note that up to this point they had engaged only in traditional agricultural activities. Not until the 1977–78 agricultural year did they begin to cultivate a nontraditional crop, tomatoes. In addition they were given calves by the absentee landowner to fatten *en compañía* on their El Queremal land.

In 1978, with the earnings from their *en compañía* crops and cows, Favio and Bolívar were able to buy 12 acres of land in El Palmar. The land, uncultivated except for .5 acre of coffee, and a large but old wooden house cost them 335,000 pesos, or 28,000 pesos per acre. They paid

115,000 pesos down, with the rest due in two years. To help finance the purchase, Favio borrowed 50,000 pesos from the Caja Agraria. In 1978 they also each received their 2.5-acre inheritances from Lubino. Their total holdings amounted to 46 acres each, 8.5 each in El Palmar. Favio was twenty-seven, Bolívar thirty-three. Favio had married the year before, and he and his wife were expecting their first child. Bolívar set up household in 1978 with a young woman from Dagua. In 1978–79 the brothers continued to grow traditional foodstuffs and tomatoes, working *en compañía* with the absentee landowner (for the last time) and with a neighbor, and beginning to cultivate their joint holding. They also began to plant new coffee (but not *caturra*) on that land. With combined farm expenses of some 35,000 pesos, the two brothers earned a combined net income of 120,000 pesos from their agricultural activities in 1978–79—more than adequate for the basic needs of the two small households.

We have looked at examples of successful households in El Palmar. Let us look at two cases of households who were doing poorly in 1978–79.

### *Case 6: The Young* Jornalero

In 1978–79 Fernando Gonzales was thirty years old. The son of a local resident with 6 acres of land, Fernando himself owned no land. His wife's father, Ramón Ortiz, who had arrived in El Palmar in 1958, owned no land, although for some time he had been renting 6 acres. The couple had five children: a thirteen-year-old daughter, a nine-year-old son, two young children, and an infant. Fernando worked as a *jornalero* on a large farm in a neighboring community, earning 450 pesos per week plus meals. His cash income in 1978–79 of 23,000 pesos was far from adequate for the needs of his large young family. His father-in-law helped them out with weekly contributions of meat (Ramón worked on weekends for one of the local butchers), without which the family's diet would have been extremely inadequate. As it was, they often had to borrow from neighbors or buy from the grocery store on credit. The family lived in a *ranchito*, a tiny mud-and-bamboo shack with a dirt floor. They had virtually no household possessions except for threadbare clothing and a few items of furniture.

### *Case 7: The Household Relying on Mixed Income Sources*

Carlos Cabrera and his family were in a similar position. He and his wife Excelinia had returned to El Palmar in 1974 with no savings and six young children. Aristides Benavides had given him a *ranchito* to live in, and

Carlos had begun to work *en compania* with him tending his coffee. But again he had not been able to save anything. The trees had been old and hard to pick, and although Carlos' sons had helped, he had had to hire workers for the harvest. Carlos had paid almost all the costs of maintaining, harvesting, and processing the coffee (Aristides paid half the cost of the weeding, or *limpieza*), and had paid Aristides half the harvest.

By 1978 Carlos was forty-six and Excelinia was thirty-three. They had seven children: two boys, aged seventeen and fourteen, and five daughters ranging in age from a toddler up to twelve. The two sons had dropped out of elementary school after a few years in order to work full-time on the farm and as *jornaleros*, while the oldest daughter had also stopped school in order to help her mother with the housework and child care. Carlos was working full-time on the 2.5 acres he had just received from Lubino as his inheritance, and on 7.5 acres of land he was working *en companía* with an absentee landowner. The land he had inherited was in old coffee and overgrown with brush, and Carlos was gradually clearing it and planting new coffee (a traditional variety, without fertilizer). On his *en companía* land he was cultivating 3 acres of maize, beans, and *arracacha*—he did not grow tomatoes or green peppers because he reacted badly to the pesticides and thought those crops risky in any case. In addition, he and his family worked 4.5 acres on that farm, and 3 acres on Aristides Benavides' farm, in coffee. In both cases the coffee trees were old and poorly tended and the yield was low. The family also supplemented their income with a variety of minor activities. Carlos bought *arracacha* from his brother Leonardo and resold it in Cali. Excelinia worked at home embroidering blouses from Cali on a piecework basis. She and her oldest daughter also took in washing. The family lived in a small three-room *ranchito* that they had built in 1977 on a lot by the highway, given to them by Lubino. Excelinia and the children sold fruits and vegetables on a commission basis from a small stall they had set up in the yard. Excelinia also raised fowl for sale.

With all of these activities the family's net income in 1978–79 was only approximately 55,000 pesos, including the value of farm produce consumed by the household. Almost the entire household income was spent on food, yet it was still below the basic subsistence requirement for this family of nine (66,400 pesos). The family had no savings and no further prospect of inheritance. Aside from the fruit and vegetable stall, they had not taken advantage of the new opportunities of the 1970s. Since he did not yet have title to his land, Carlos had been unable to take out a loan to plant new coffee and had been unwilling to plant tomatoes or green peppers. He had never borrowed from a commercial source, and his only purchased inputs were chicken manure (*gallinaza*) and hired labor for his *en companía* crops. Unless Carlos changed his strategy, it was unlikely that the family's position would improve in the near future.

## Economic Status from 1978–79 to 1983–84

### *Case 8: Farmers Specializing in Coffee*

Farmers like Arcesio Benavides and his wife Diomelina, who had been concentrating on coffee production in 1978–79, had maintained or improved their material conditions in this period. Arcesio and Diomelina had constructed a third bedroom and extended the drying roof. Arcesio's land now had 1.5 acres in sugar cane and 4.5 acres in coffee. Three acres of that was in *caturra*, financed by bank loans. Diomelina had taken out loans to purchase 3.75 more acres of land, bringing her total holdings to 7.5 acres. Of that, she had given 4.5 acres to her oldest son (1.5 in *caturra*), who now lived with his wife and child in the old Álvarez family house. Diomelina had given another .75 acre to her oldest daughter, who lived in Cali. The daughter had followed her brother's lead by taking out a loan to raise honey bees. Diomelina had secured that loan with her land. Another .75 acre was used to pasture the horse used for all the family's operations. Diomelina herself operated 1.5 acres of her land in new *caturra*, and another 1.5 acres of old coffee on a brother-in-law's land *en companía*. The income from her operations had gone to repay part of the loans for home and farm improvements, and to pay for part of the education expenses of her two youngest children.

The combined debts of Arcesio and Diomelina in December 1983 amounted to 168,000 pesos plus interest (approximately 35,000 pesos), all of which was due in 1984, but they only had 22,000 pesos available for repayment. Their oldest daughter also had large debts, for which Diomelina was ultimately responsible. The family's cash shortage occurred because 1983 had been a disastrous year for coffee: the area of El Palmar had been hit first by a frost and then by a severe drought, with the consequence that the just-completed harvest had been at a record low level. Their indebtedness made them particularly vulnerable to these natural disasters. Although the family's income level and material standard of living had been high since 1978–79, this had been possible only through continued high levels of indebtedness and a fairly complex process of debt juggling.

They planned to cover their current debt with new loans. Each new or renegotiated loan increased their interest payments, however, which they were finding extremely burdensome. Ultimately, they hoped to be able to pay off their debts with earnings from their coffee, *panela*, and some pigs that Diomelina was raising. (As a postscript, their income in 1984 was not sufficient to pay off their debts, and Diomelina obtained new loans, which she used to cover some medical expenses and make interest payments. By 1985 creditors began to threaten Diomelina with loss of her land if she did not begin to repay her daughter's debts. She was feeling increasingly desperate and finally, in 1986, sold some of her land to settle her debts.)

### *Case 9: Farmers Switching from Diversification to Coffee Specialization*

Another household in a similar condition to that of Arcesio and Diomelina in 1983 was the family of Enrique and Marina Valencia. In 1978–79 Enrique had had a 10.5-acre diversified farm. He had grown coffee, plantains, tomatoes, green peppers, *arracacha*, string beans, maize, and beans, for the market and household consumption, and a variety of fruits and herbs for household use. Only 2.25 acres of coffee had been in production, but he had also had 5.25 acres recently planted to *caturra*. Two grown sons had been working full-time with their father, and a daughter-in-law and two younger daughters had been helping with household and farm chores. The family had had one permanent worker but used no day laborers. The farm had had high outlays on seed and on fertilizer and other chemicals. Although expenses had been high, farm income was also high, and net income (130,000 pesos) had been far above household needs. As with Arcesio and Diomelina, the Valencia household had been heavily in debt in 1978–79. They had borrowed for the *caturra*, and also to begin home and farm improvements similar to those of Arcesio and Diomelina (new house and kitchen, drying roof, *beneficiadero*, as well as a TV, refrigerator, furniture, and sewing machine).

By 1979 the new coffee had begun to produce, and labor needs shot up. The family hired workers for the coffee harvest and began to cut back on other agricultural activities. Tomatoes and green peppers were dropped because they needed too much attention. As coffee yields increased over the next several years, the household did very well. They repaid most of the coffee loan and a total of 250,000 pesos for the home and farm improvements, plus interest. All the earnings of the farm went to cover farm expenses, household expenses, and loan repayment, however, and they had no savings. In 1982 they had a very good year and entered 1983 with cash on hand. But they, too, were affected by the adverse weather conditions in 1983. By October they had to borrow to cover farm and household expenses, and again in December. By the end of the year they owed 95,000 pesos, due the next year. However, they expected the next harvest (the *traviesa*) to be abundant and cover their debt and expenses.

Enrique had also decided to take a major step: he had borrowed 500,000 pesos from the Banco Cafetero (at 20 percent interest) to plant 5 acres of pineapples on land owned by his uncle, Israel Valencia. The Coffee Federation had been encouraging this activity, and Enrique had decided to take the chance because he did not think coffee would continue to be lucrative. Since his own land was almost all in fairly new coffee, he had arranged to grow the pineapples *en companía* with his uncle. In 1984 interest and principal payments on the loan would be 199,000 pesos, which he hoped to repay with coffee earnings (the pineapples would not begin producing until 1985). As was the case with Arcesio and Diomelina, then, this family had enjoyed a prosperous period and had much to show for it in home and farm improvements, but had been able to sustain their high standard of living only by placing themselves in a precarious debt situation. Enrique and Marina were talking seriously of selling their farm in El Palmar and moving to a more isolated area, where they could acquire more land and grow more foodstuffs. (They, and other members of the Evangelical sect to which they belonged, were convinced that nuclear war was coming and that this would be their best chance of survival.)

### *Case 10: Farmers with Diversified Operations*

Two farmers who had done particularly well were Favio and Bolívar Cabrera. These brothers had maintained diversified operations including maize and beans, *arracacha*, tomatoes, green peppers, dairy cows, pigs, and six acres in *caturra*. Although they borrowed every year from the Caja Agraria for operating expenses, Bolívar, who was once again single, had accumulated savings, which he kept in the Banco Cafetero (where, according to Bolívar, it earned 33 percent interest, compared to the 30 percent they paid the Caja for working capital). Bolívar had also begun to lend money to neighbors at 5 percent monthly interest.

### *Case 11:* Jornaleros

Fernando Gonzales had continued to work as an agricultural day laborer on a large neighboring estate. In addition, he worked *en companía* with his father-in-law on a small plot of land that Ramón had purchased. Fernando's thirteen-year-old son also worked on Ramón's plot when he was not in school. Ramón, who ate with them, provided the household with eight to ten pounds of meat weekly from his work butchering cows. The family had moved from the *ranchito* to a larger, better-constructed house. Their cash income of 55,000 pesos in 1983 was supplemented by approximately 80,000 pesos' worth of meat and *en companía* produce, so that overall the family's earnings were more than sufficient for their subsistence needs. Fernando's wife, Luz, had bought a sewing machine with a loan from her godmother, which she was repaying with earnings from raising chickens and from some work picking coffee. Although the family's material conditions had improved, however, they were clearly not in a position to acquire farmland. In 1983 their situation was threatened by a decision of Fernando's employer to use labor contractors (*contratistas*, hired to carry out specific work projects, who in turn hire the necessary labor) rather than continue to hire his labor force directly. As a result, Fernando's employment on the estate was no longer regular.

### *Cases 12 and 13:* Agregados/Mayordomos

Two of the *agregado/mayordomo* households we looked at earlier had also done well in the intervening years. In chapter 5 we saw that Luis Murcia and his wife Carmen had gone through a difficult period in the 1960s. In the 1970s the family had become *agregados* for the large absentee-owned farm discussed in chapter 6 (the same farm where the Cabrero brothers worked *en companía*). Luis had received a small salary and *en companía* income and products from the farm's crops, coffee, and dairy operations. By 1977 Luis had been able to buy a small plot of land by the highway (from the absentee owner), where they had begun to build a house. Luis' two oldest sons (aged twenty and fifteen) had worked full-time with him in 1978–79; his wife, daughter (aged eighteen), and a younger son (aged eleven) had carried out many of the home and farm chores. With two young children in addition, the large household had had high food requirements, and their net income had not been much over subsistence level (a ratio of 1:2). Their income increased over the next several years, however. When the absentee owner died, they received a generous settlement from the heirs, and Luis was hired at a good salary by the new owner. But Luis wanted his independence, and soon quit the farm to open a *cantina* in the market. They were pleased with their economic situation in 1983. The *cantina* was doing well. Their second-oldest son, now twenty, had been able to return to school and was attending the high school in Bitaco.

We saw above that the *mayordomo* Omar Gomez and his wife Marta were doing well in 1978–79 and hoped to soon acquire land in El Palmar. This they were able to do in 1981, buying a 9.75 acre farm for 600,000 pesos. They built a house on their farm, and in 1983 had a family living there and workers and an administrator on their farm, while they continued to live and work on Emilio Quintero's farm. They and their young child were in a very favorable economic position.

Although other *agregados* and *mayordomos* in El Palmar were not doing as well as Omar, they, too, had generally improved their circumstances since 1978–79. Many had been able to buy small plots of land even at the elevated prices of the period.

### *Cases 14 and 15: Shopkeepers*

The period between 1978–79 and late 1983 had seen a proliferation of small shops along the highway and the roads of the community: grocery stores, food kiosks, bars, and dance halls abounded. Some of the earlier shopkeepers felt that their business had not been affected by the increased competition. Some of the newer entrants, like Luis and Carmen Murcia, found their business profitable. Some earlier shopkeepers had been able to expand since 1978–79. This was the case with Gabriela Diaz and her husband Claver Quintero, whom we last looked at in 1968, when they had finally managed to buy 1.5 acres of land to supplement Gabriela's 1.5 acre inheritance (chapter 4). In 1974 they had purchased a tiny lot by the highway, where Gabriela had set up a small food kiosk, selling *empanadas*, hard-boiled eggs, and other snacks to passersby and to the weekend evening party-goers. In 1978 they had enlarged the lot slightly with another purchase so that Gabriela could expand her *negocio*. On the farm, Claver had taken out a small loan from the Coffee Cooperative and eliminated all other crops to plant their holdings entirely in coffee. He had divided his time in 1978–79 between work on the farm and agricultural labor. His oldest son (aged twenty-three) had worked on the farm. The two oldest daughters had left home, the older to marry, the younger at the age of fifteen to work as a domestic in Cali. The next daughter (aged seventeen) had helped Gabriela, and the two youngest children had been in school. The family's 1978–79 net income of 74,000 pesos from the farm, agricultural labor, and Gabriela's *negocio* had been almost double their subsistence requirements; half of that income had come from the *negocio*.

By 1983 their *negocio* had expanded. In order to pay their debt for the *negocio* lot, Gabriela and Claver had sold half their farm to their oldest son, keeping 1.5 acres in new coffee. With income from their holding, from Claver's work off the farm, and from Gabriela's business, they had been able to build a small store with a kitchen and living quarters, where they moved with their two youngest children. The daughter was attending high school in Bitaco in 1983. The son, eighteen, helped on the family's small plot and worked as an agricultural laborer. Claver worked part-time on his holding and was employed part-time by the community as the *montanero*, maintaining the aqueduct. The family's 1983 income for four people was 150,000 pesos, more than double their 1978–79 income per family member. They felt that their income was more than sufficient for their needs, even with rising food costs and a daughter in high school (school expenses, including matriculation, uniforms, books, meals, and transportation, amounted to over 27,000 pesos per year), and that their economic situation had improved.

Not all the original shopkeepers felt unaffected by the increased competition of the early 1980s. Norberto Bermúdez, in particular, had found that his *cantina* business had fallen off considerably. By 1982 he and his family were drawing on the interest from their savings, and by 1983 beginning to live off the savings themselves. Their net income in 1983 from the shop was only around 50,000 pesos, half of the 1978–79 figure, while their household expenses had increased. Norberto had also invested in a business in Felidia that was going badly. He and Nidia were thinking of selling out and moving to Cali with their two children. In general, however, the Bermúdez family appeared to be an exception among the families with shops, most of whom felt that conditions had been favorable.

### *Case 16: Traders*

Middlemen like Desiderio Almeida appeared in general to be prospering despite increased competition. Although other buyers had entered the resale business, there appeared to be enough growers to maintain an increased number of traders. Desiderio continued to work with the same growers he had dealt with for many years. As in 1978–79, much of his income was tied up in advances to these growers. He had had some setbacks. In 1983 he had exchanged 6 acres of land for a pickup truck that had been too small to be of much use in his business, and he had ultimately sold it at a loss. His trading was risky. Desiderio traveled to Cali only once a week to market produce, purchase inputs, and assess prices. Based on those prices, he determined the prices he would offer his growers during the following week. However, the Cali prices fluctuated frequently, and he sometimes received less in Cali than he had paid. On the whole, however, the trading had been profitable. Desiderio and his family were in a comfortable position with his farm and his business.

# References

Abbott, Joan. 1974. "The Employment of Women and the Reduction of Fertility: Implications for Development." *World Development* 2, no. 2.

Adams, Dale W., and S. Schulman. 1967. "Minifundia in Agrarian Reform: A Colombian Example." *Land Economics* 43, no. 3.

Afshar, Haleh. 1985. "The Position of Women in an Iranian Village." In Haleh Afshar, ed., *Women, Work, and Ideology in the Third World.* London: Tavistock.

Amin, Samir. 1974. "Capitalism and Ground Rent: The Domination of Capitalism over Agriculture in Tropical Africa." Mimeographed. Dakar: United Nations African Institute for Economic Development and Planning.

Amin, Samir, and Kostas Vergopoulos. 1974. *La Question paysanne et le capitalisme*. Paris: Anthropos. Available in Spanish as *La cuestión campesina y el capitalismo*. Mexico: Editorial Nuestro Tiempo, 1975.

Anker, Richard, Mayra Buvinic, and Nadia H. Youssef, eds. 1982. *Women's Roles and Population Trends in the Third World*. Geneva: International Labour Organization.

Arango Londoño, Gilberto. 1972. *Estructura económica colombiana*. Bogotá: Banco del Comercio.

Arrubla, Mario, et al. 1978. *Colombia: Hoy.* Bogotá: Siglo Veintiuno Editores de Colombia.

Arrubla, Mario, ed. 1976. *La agricultura colombiana en el siglo XX*. Bogotá: Editorial Andes.

Asad, Talal, and Harold Wolpe. 1976. "Concepts of Modes of Production." *Economy and Society* 5, no. 4.

Asociación Colombiana de Ganaderos. 1958. "Industria Agropecuaria-CVC-Monocultivo." Report by the president of Adega-Valle, Dr.

Camilo Molina Ossa, to the Asambléa General de Ganaderos, 27 October.
Banaji, Jairus. 1977. "Modes of Production in a Materialist Conception of History." *Capital and Class*, no. 3.
Banco de la República. 1983. *Revista Banco de la República* (Bogotá), July.
Banderas, Pedro Antonio. 1944. *Diccionario geográfico industrial y agrícola del Valle del Cauca*. Cali: n.p.
Barraclough, Solon, ed. 1973. *Agrarian Structure in Latin America: A Resume of the CIDA Land Tenure Studies*. Lexington, Mass.: D. C. Heath.
Barraclough, Solon L., and Arthur L. Domike. 1966. "Agrarian Structure in Seven Latin American Countries." *Land Economics* 42, no. 4. Reprinted in Rodolfo Stavenhagen, ed., *Agrarian Problems and Peasant Movements in Latin America*. New York: Anchor Books, 1970.
Bartra, Roger. 1974. *Estructura agraria y clases sociales en México*. Mexico City: Editorial ERA.
———. 1975. "La teoría del valor y la economía campesina." *Comercio Exterior* 25, no. 5.
———. 1976. "Si los campesinos se extinguen." *Historia y Sociedad*, no. 8.
Becerra Collazos, Guillermo. 1977. *Monografía de Dagua*. Dagua: n.p.
Becker, Gary S. 1960. "An Economic Analysis of Fertility." In NBER, *Demographic and Economic Change in Developed Countries*. Princeton, N.J.: National Bureau of Economic Research.
———. 1965. "A Theory of the Allocation of Time." *Economic Journal* 75 (Spring).
Becker, Gary, and H. G. Lewis. 1974. "Interaction between Quantity and Quality of Children." In T. W. Schultz, ed., *Economics of the Family*.
Beechey, Veronica. 1977. "Some Notes on Female Wage Labor in the Capitalist Mode of Production." *Capital and Class*, no. 3.
———. 1979. "On Patriarchy." *Feminist Review*, no. 3.
Bejarano, Jesús A. 1975. "El fín de la economía exportadora y los orígenes del problema agrario." *Cuadernos Colombianos* (Medellín: Editorial Lealon), nos. 6–8.
———. 1978. "Industrialización y política económica, 1950–1976." In Mario Arrubla et al., *Colombia*.
———. 1979. *El régimen agrario: De la economía exportadora a la economía industrial*. Bogotá: La Carreta.
Beneria, Lourdes, ed. 1982. *Women and Development*. New York: Praeger.
Beneria, Lourdes, and Gita Sen. 1981. "Accumulation, Reproduction,

and Women's Role in Economic Development: Boserup Revisited." *Signs* 7, no. 2.

———. 1982. "Class and Gender Inequalities and Women's Role in Economic Development—Theoretical and Practical Implications." *Feminist Studies* 8, no. 1.

Bergquist, Charles. 1986. *Labor in Latin America.* Stanford, Calif.: Stanford University Press.

Bernstein, Henry. 1977. "Notes on Capital and Peasantry." *Review of African Political Economy,* no. 10.

———. 1979. "African Peasantries: A Theoretical Framework." *Journal of Peasant Studies* 6, no. 4.

Berry, Albert, and William Cline. 1979. *Agrarian Structure and Productivity in Developing Countries.* Baltimore: Johns Hopkins University Press.

Birdsall, Nancy. 1976. "Women and Population Studies." *Signs* 1, no. 2.

Bonilla, Victor Daniel. 1967. "El Valle del Cauca a la hora de su transformación agraria." *Tierra* (Bogotá: Ediciones Tercer Mundo), no. 6.

Boserup, Esther. 1970. *Woman's Role in Economic Development.* New York: St. Martin's Press.

Bourque, Susan, and Kay Warren. 1981. *Women of the Andes: Patriarchy and Social Change in Two Peruvian Towns.* Ann Arbor: University of Michigan Press.

Bradby, Barbara. 1975. "The Destruction of Natural Economy." *Economy and Society* 4, no. 2.

Braverman, Harry. 1974. *Labor and Monopoly Capital.* New York: Monthly Review Press.

Brenner, Johanna, and Maria Ramas. 1984. "Rethinking Women's Oppression." *New Left Review,* no. 144.

Brewster, John M. 1950. "The Machine Process in Agriculture and Industry." *Journal of Farm Economics* 32 (February–August).

Bryceson, Deborah. 1983. "Use Values, the Law of Value and the Analysis of Non-Capitalist Production." *Capital and Class,* no. 20.

Bulatao, R. A. 1978. *The Value of Children: A Cross National Study: The Philippines.* Honolulu: East-West Population Institute.

Burawoy, Michael. 1976. "The Functions and Reproduction of Migrant Labour: Comparative Material from Southern Africa and the United States." *American Journal of Sociology* 81, no. 5.

Cáceres G., Ingrid. 1980. "La división de trabajo por sexo en la unidad campesina minifundista." In León de Leal and Deere, *Mujer y capitalismo agrario.*

Caldwell, John. 1982. *Theory of Fertility Decline.* London: Academic Press.

Camacho Peréa, Miguel. 1962. *El Valle del Cauca, constante socioeconómica de Colombia.* Cali: Imprenta Departamental.

CENAGRO [Comité Regional de Producción Agrícola del Valle del Cauca]. 1978. *Censo agropecuario del Valle del Cauca.* Cali: CENAGRO.

Chayanov, Alexander Vasilevich. 1966. "Peasant Farm Organization" (Organizatsiya krest'yanskogo khozyaistiva). In D. Thorner, B. Kerblay, and R. E. F. Smith, eds., *The Theory of Peasant Economy.* Homewood, Ill.: Richard D. Irwin.

CIDA [Inter-American Committee for Agricultural Development]. 1966. *Tenencia de la tierra y desarrollo socio-económico del sector agrícola: Colombia.* Washington, D.C.: CIDA.

———. 1967. *Una experiencia en diversificación de cultivos y desarrollo regional—Caldas, Colombia.* Washington, D.C.: CIDA.

CIEP. 1979. "Cae Precio Interno del Cafe." *El País,* 26 February.

Clawson, David, and Don Hoy. 1979. "Nealticon, Mexico: A Peasant Community That Rejected the 'Green Revolution.'" *American Journal of Economics and Sociology* 38, no. 4.

Cleaver, Harry. 1972. "The Contradictions of the Green Revolution." *American Economic Review* 62 (May).

Colmenares, Germán. 1976. *Cali: Terratenientes, mineros y comerciantes, siglo XVIII.* Cali: Universidad del Valle.

Corchuelo, Alberto. 1979. "Condiciones de desarrollo de la agricultura y la política agraria." *Cuadernos Colombianos,* no. 12.

Cordtz, Dan. 1972. "Corporate Farming: A Tough Row to Hoe." *Fortune,* August.

Crener, Maxime A., Gaston Léal, Robert LeBlanc, and Brigitte Thébaud. 1984. *Integrated Rural Development: State of the Art Review—1982/83.* Ottawa: IDRC.

Crist, Raymond E. 1952. *The Cauca Valley, Colombia: Land Tenure and Land Use.* Baltimore: Waverly Press.

Crittenden, Ann. 1981. "In Colombia, Science Sows Seeds of Hope for the Hungry." *New York Times,* 17 May.

Crouch, Luis, and Alain de Janvry. 1980. "The Class Basis of Agricultural Growth." *Food Policy* 5, no. 1.

Currie, Lauchlin. 1950. *The Basis of a Development Program for Colombia.* Report of a mission headed by Lauchlin Currie and sponsored by the International Bank for Reconstruction and Development in collaboration with the Government of Colombia. Washington, D.C.: IBRD.

CVC [Corporación Autónoma Regional del Valle del Cauca]. 1971. *Cuenca superior del Río Nima: Inventorio del uso actual y tenencia de la tierra.* CVC Report no. 71–14. Cali: CVC.

———. 1975. *El Valle del Cauca en la Economía Nacional.* Report CVC no. 75–19. Cali: CVC, Departamento de Estudios Económicos.

———. 1976. *Plan de ordenación y desarrollo de las cuencas de los Ríos Jamundí, Claro y Timba.* CVC Report no. 76–3. Cali: CVC.

———. 1977a. *Plan de ordenación y desarrollo de las cuencas de los Ríos Bolo-Fraile-Desbaratado.* CVC Report no. 77–5. Cali: CVC.

———. 1977b. *Valle del Cauca: Manual de costos de producción agrícola.* CVC Report no. 77–16. Cali: CVC.

———. 1978. *El Valle del Cauca en la Economía Nacional, 1977.* CVC Report no. 78–1. Cali: CVC.

DANE [Departamento Administrativo Nacional de Estadística]. N.d. Colombian population censuses, 1825, 1835, 1843, 1851, 1864, 1870, 1905, and 1912. Untitled bound volume. Cali: DANE, Office of Population Census.

———. 1942. *Censo general de la población, 5 de julio de 1938.* Vol. 14, *Departamento del Valle del Cauca.* Bogotá: Imprenta Nacional. Also available in bound volume, DANE, Office of Population Census, Cali, n.d.

———. 1954. *Muestra agropecuaria nacional, 1954.* Bogotá: DANE.

———. 1955. *Panorama geo-económico del departamento del Valle.* Bogotá: DANE.

———. 1962. *Censo agropecuario nacional 1960.* Bogotá: DANE.

———. 1970. "La reforma agraria en cifras." *Boletín Mensual de Estadística,* no. 222.

———. 1973a. *Censo nacional agropecuario, 1970–71.* Bogotá: DANE, División de Estudios Económicos.

———. 1973b. *XIV censo nacional de población y III de vivienda.* Bogotá: DANE.

———. 1975. *Estadísticas históricas.* Bogotá: DANE.

———. 1978. *Boletín mensual de estadística,* no. 324.

DAP [Departamento Administrativo de Planeación]. 1978. *Anuario estadístico del Valle del Cauca, 1976.* Cali: DAP.

Deere, Carmen Diana. 1976. "Rural Women's Subsistence Production in the Periphery." *Review of Radical Political Economics* 8, no. 1.

———. 1977. "Changing Social Relations of Production and Peruvian Peasant Women's Work." *Latin American Perspectives* 4, nos. 1 and 2.

———. 1988. "Household and Class Relations: The Peasant Economy of the Northern Peruvian Highlands, 1900–1980." Manuscript. Amherst: Department of Economics, University of Massachusetts.

Deere, Carmen Diana, and Alain de Janvry. 1979. "A Conceptual Framework for the Empirical Analysis of Peasants." *American Journal of Agricultural Economics* 61, no. 4.

Deere, Carmen Diana, and Magdalena León de Leal. 1982. "Peasant Production, Proletarization, and the Sexual Division of Labor in the Andes." In Beneria, ed., *Women and Development*.

Deere, Carmen Diana, Peter Marchetti, and Nola Reinhardt. 1985. "The Peasantry and the Development of Sandinista Agrarian Policy, 1979–1984." *Latin American Research Review* 20, no. 3.

De Janvry, Alain. 1981. *The Agrarian Question and Reformism in Latin America*. Baltimore: Johns Hopkins University Press.

De Janvry, Alain, and Carlos Garramon. 1977. "The Dynamics of Rural Poverty in Latin America." *Journal of Peasant Studies*, no. 4.

Delphy, Christine. 1977. *The Main Enemy: A Materialist Analysis of Women's Oppression*. London: Women's Research and Resource Center.

Den Hartog, A. P. 1973. "Unequal Distribution of Food within the Household." *FAO Newsletter* 10, no. 4.

Diaz, Antolín. 1935. *Sinú, pasión y vida del trópico*. Bogotá: Editorial Santa Fe.

Diaz del Castillo, Porfirio. 1937. *El Valle del Cauca, historia y realidad de sus municipios*. Cali: Imprenta Marquez.

Dirección General DRI–PAN. 1982a. *Los programas de pancoger y boletos y la reubicación geográfica del DRI: Relación con las nuevas orientaciones del programa DRI/PAN*. Bogotá: Departamento Nacional de Planeación.

———. 1982b. *Realizaciones de los programas DRI y PAN*. Bogotá: Departamento Nacional de Planeación.

División de Evaluación PAN–DRI. 1982. "Informe de ejecución acumulada, 1981." Bogotá: Departamento Nacional de Planeación.

Dorner, Peter, and Don Kanel. 1971. "The Economic Case for Land Reform: Employment, Income Distribution, and Productivity." In Peter Dorner, ed., *Land Reform in Latin America: Issues and Cases*. Madison, Wis.: Land Economics.

Dupre, Georges, and Pierre Philippe Rey. 1973. "Reflections on the Pertinence of a Theory of the History of Exchange." *Economy and Society* 2, no. 2.

Duque, Germán. 1984. "Hacia donde va el DRI?" *Nueva Frontera* (Bogotá), 6 August.

Durrenberger, Paul E., ed. 1984. *Chayanov, Peasants and Economic Anthropology*. New York: Academic Press.

Echavarría U., Juan Fernando. 1976. "Contribución al análisis del sector agraria: El problema de la forma de producción parcelaria." *Revista de Extensión Cultural* (Medellín: Universidad Nacional), nos. 2–3.

ECLA [United Nations Economic Commission for Latin America]. 1956.

*Analyses and Projections of Economic Development. 3, The Economic Development of Colombia.* E/CN.12/365/Rev.1. New York: United Nations.

———. 1975. "Participation of Women in Development in Latin America." United Nations World Conference of the International Women's Year. Conference background paper no. E.Conf.66/BP/8/Add1. New York: ECLA.

———. 1983. *Five Studies on the Situation of Women in Latin America.* Santiago, Chile: CEPAL.

Economist Intelligence Unit. 1962–73. *Quarterly Economic Review: Colombia.* London: Spencer House.

Easterlin, R. A. 1975. "An Economic Framework for Fertility Analysis." *Studies in Family Planning* 5 (March).

———, ed. 1980. *Population and Economic Change in Developing Countries.* Chicago: University of Chicago Press.

Edholm, Felicity, Olivia Harris, and Kate Young. 1977. "Conceptualizing Women." *Critique of Anthropology* 3, nos. 9–10.

Eisenstein, Zillah R., ed. 1979. *Capitalist Patriarchy and the Case for Socialist Feminism.* New York: Monthly Review Press.

*El Siglo.* 1983. "Ampliarán diversificación en zonas cafeteras." *El Siglo,* 11 November.

———. 1984. "El DRI: Nuevo instituto de desarrollo rural." *El Siglo,* 21 January.

Ennew, Judith, Paul Hirst, and Keith Tribe. 1977. "'Peasantry' as an Economic Category." *Journal of Peasant Studies* 4, no. 4.

Escorcia, José. 1978. *Historia económica y social de Colombia, siglo XX.* Bogotá: Editorial Presencia.

———. 1980. "La formación de las clases sociales en una sociedad multiétnica: Cali, 1820–1854." *Historia y espacio: Revista de Estudios Históricos Regionales* (Colombia) 2, nos. 6–7.

———. 1983. *Desarrollo político, social y económico, 1800–1854.* Bogotá: Fondo de Promoción de la Cultura del Banco Popular.

Espenshade, Thomas. 1977. "The Value and Cost of Children." *Population Bulletin* 32, no. 1.

Esteva, Gustavo. 1978. "Y si los campesinos existen?" *Comercio Exterior* (Mexico City), no. 28.

Fajardo, Darío M. 1983. *Haciendas, campesinas y políticas agrarias en Colombia, 1920–1980.* Bogotá: Editorial La Oveja Negra.

Fajardo, Darío M., Absalón Machado, Piedad Gómez Villa, María Salazar, León Zamosc, María Gaitán de Pombo, and W. Ramírez Tobón. 1981. *Campesinado y capitalismo en Colombia.* Bogotá: CINEP.

Falcon, Walter P. 1970. "The Green Revolution: Generations of Problems." *American Journal of Agricultural Economics* 52, no. 5.

Fals Borda, Orlando. 1955. *Peasant Society in the Colombian Andes: A Sociological Study of Saucio*. Gainesville: University of Florida Press. Available in Spanish as *Campesinos de los Andes: Estudio sociológico de Saucío*. Monografías Sociológicas, no. 7. Bogotá: Universidad Nacional, 1971.

———. 1957. *El hombre y la tierra en Boyacá: Bases sociológicas y históricas para una reforma agraria*. Bogotá: Ediciones Documentos Colombianos.

———. 1969. *Subversion and Social Change in Colombia*. New York: Columbia University Press.

FAO [United Nations Food and Agriculture Organization]. 1947. *The World's Coffee*. Rome: FAO. Also available in Spanish as *El café en el mundo*.

FEDECAFÉ [Federación Nacional de Cafeteros Colombianos]. 1963. *Manual de instructivos agro-técnicos para el proyecto de crédito rural planeado (para Caldas)*. FEDECAFÉ.

———. 1975. *Campañas técnicas 1967–1970: Anexo del informe de la gerencia al XXIX Congreso Nacional de Cafeteros*. Cali: Biblioteca del Comité Cafetero.

———. 1977. *Programa de desarrollo y diversificación de zonas cafeteras*. FEDECAFÉ.

———. 1978. *Informe de gerente general*. Bogotá: FEDECAFÉ.

Feder, Ernest. 1971. *The Rape of the Peasantry: Latin America's Landholding System*. New York: Doubleday.

———. 1976. "The New World Bank Program for the Self-Liquidation of the Third World Peasantry." *Journal of Peasant Studies* 3, no. 3.

———. 1977. "Agribusiness and the Elimination of Latin America's Rural Proletariat." *World Development* 5, nos. 5–7.

FEDESARROLLO. 1978. *Economía cafetera colombiana*. Bogotá: Editorial Andes.

Felstehausen, Herbert. 1971. "Agrarian Reform: Colombia." In Peter Dorner, ed., *Land Reform in Latin America: Issues and Cases*. Madison, Wis.: Land Economics.

Fogel, Robert, and Stanley Engerman. 1974. *Time on the Cross: The Economics of American Negro Slavery*. Boston: Little, Brown.

Folbre, Nancy. 1982. "Exploitation Comes Home: A Critique of the Marxian Theory of Family Labour." *Cambridge Journal of Economics* 6, no. 4.

———. 1983. "Of Patriarchy Born: The Political Economy of Fertility Decisions." *Feminist Studies* 9, no. 2.

———. 1984. "A Patriarchal Mode of Production." Mimeographed. Amherst: Department of Economics, University of Massachusetts.

———. 1985. "The Wealth of Patriarchs: Deerfield, Massachusetts, 1760–1840." *Journal of Interdisciplinary History* 16, no. 2.

———. 1986. "Cleaning House: New Perspectives on Households and Economic Development." *Journal of Development Economics* 22, no. 1.

Forero Álvarez, Jaime, and Guillermo Rudas. 1983. *Producción y comercialización de perecederos agrícolas en zonas campesinas integradas al mercado de Bogotá*. Bogotá: Pontificia Universidad Javeriana.

Foster, George M. 1967. "Introduction: What Is a Peasant?" In Potter, Foster, and Diaz, eds., *Peasant Society*.

Franco Mejía, Lilia Evelyn. 1980. "La historia de Buenaventura y su área de influencia." Thesis. Cali: Universidad del Valle.

Friedmann, Harriet. 1978. "Simple Commodity Production and Wage Labour in the American Plains." *Journal of Peasant Studies* 6, no. 1.

———. 1980. "Household Production and the National Economy: Concepts for the Analysis of Agrarian Formations." *Journal of Peasant Studies* 7, no. 2.

Gardiner, Jean. 1975. "Women's Domestic Labour." *New Left Review*, no. 89.

Gilhodes, Pierre. 1970. "Agrarian Struggles in Colombia." In Stavenhagen, ed., *Agrarian Problems and Peasant Movements*. Available in Spanish as *Las luchas agrarias en Colombia*. Medellín: Editorial Lealon, 1976.

Gomez, Fernando. 1970. "Los censos en Colombia antes de 1905." In Urrutia and Arrubla, eds. *Compendio de estadísticas*.

Gomez Benítez, Piedad. 1979. "El camino de Buenaventura." Thesis. Cali: Universidad de Valle.

Gómez Villa, Piedad. 1981. "Economía campesina y descomposición del campesinado." In Fajardo et al., *Campesinado y capitalismo*.

González G., Clara. 1980. "Formación y cambios del latifundio ganadero y sus efectos en la organización de la familia campesina." In León de Leal and Deere, eds., *Mujer y capitalismo agrario*.

Goodman, David, and Michael Redclift. 1982. *From Peasant to Proletarian, Capitalist Development and Agrarian Transition*. New York: St. Martin's Press.

Griffin, Keith. 1974. *The Political Economy of Agrarian Change*. London: Macmillan.

Guyer, Jane. 1980. "Female Farming and the Evolution of Food Production Patterns amongst the Beti of South Central Cameroon." *Africa* 50, no. 4.

Guzmán Campos, Germán, Orlando Fals Borda, and Eduardo Umaña Luna. 1962, 1964. *La Violencia en Colombia: Estudio de un proceso social.* 2 vols. Universidad Nacional, Monografías Sociológicas, no. 12. Bogotá: Ediciones Tercer Mundo.

Haggag Youssef, Nadia. 1974. *Women and Work in Developing Countries.* University of California Population Monograph Series, no. 15 (Berkeley).

Haney, Emil B., Jr. 1969. "The Economic Reorganization of Minifundia in a Highland Community of Colombia." Ph.D. diss., University of Wisconsin.

Harrison, Mark. 1977. "The Peasant Mode of Production in the Work of A. V. Chayanov." *Journal of Peasant Studies* 4, no. 4.

Hartmann, Heidi. 1981a. "The Family as a Locus of Gender, Class, and Political Struggle: The Example of Housework." *Signs* 6 (Spring).

———. 1981b. "The Unhappy Marriage of Marxism and Feminism: Towards a More Progressive Union." In Lydia Sargent, ed., *Women and Revolution: A Discussion of the Unhappy Marriage of Marxism and Feminism.* Boston: South End Press.

Havens, Eugene. 1966. *Támesis: Estructura y cambio.* Bogotá: Editorial Tercer Mundo.

Henn, Jeanne. 1986. "The Material Basis of Sexism: A Mode of Production Analysis with African Examples." Boston University African Studies Center, Working Papers in African Studies, no. 119.

———. 1987. "Intra-Household Dynamics and State Policies as Constraints on Food Production: Results of a 1985 Agroeconomic Survey in Cameroon." In Susan Poats, Marianne Schmink, and Anita Spring, eds., *Gender Issues in Farming Systems Research and Extension.* Boulder, Colo.: Westview Press.

Heynig, Klaus. 1982. "The Principal Schools of Thought on the Peasant Economy." *CEPAL Review,* no. 16.

Himmelweit, Susan, and Simon Mohun. 1977. "Domestic Labour and Capital." *Cambridge Journal of Economics* 1, no. 1.

Hindess, Barry, and Paul Q. Hirst. 1975. *Pre-Capitalist Modes of Production.* London: Routledge and Kegan Paul.

———. 1977. *Mode of Production and Social Formation.* London: Macmillan.

Hirschman, Albert O. 1963. *Journeys toward Progress.* New York: Twentieth Century Fund.

Hirschon, R., ed. 1983. *Women and Property, Women as Property.* London: Croom Helm.

Ho, Teresa. 1979. "Time Costs of Childrearing in the Philippines." *Population and Development Review* 5, no. 4.

Hopper, David. 1955. "Allocative Efficiency in Traditional Indian Agriculture." *Journal of Farm Economics*, no. 47.

Hyland, Richard P. 1979. Copy of a letter to Germán Colmenares, presenting some of the results of Hyland's dissertation research in Colombia, 1977. See "The Secularization of Credit in the Cauca Valley, Colombia, 1851–1880." Ph.D. diss. University of California, Berkeley.

———. 1982. "A Fragile Prosperity: Credit and Agrarian Structure in the Cauca Valley, Colombia, 1851–1887." *Hispanic American Historical Review* 62, no. 3.

IBRD [International Bank for Reconstruction and Development]. 1956. *The Agricultural Development of Colombia*. Washington, D.C.: IBRD.

ICA [Instituto Colombiano Agropecuario]. 1972. *Estudio socio-económico del oriente antioqueño*. Bogotá: ICA.

———. 1973. *Projecto de desarrollo rural García Rovira: Estudio básico regional (informe preliminar)*. Bogotá: ICA.

IDB [International Development Bank]. 1972. *Evaluación de proyecto de desarrollo y diversificación en la zona cafetera de Caldas, Colombia*. IDB-FEDECAFÉ.

IFA [Instituto de Fomento Algodonero]. 1967. *Descripción agrícola del Valle*. IFA.

IMF [International Monetary Fund]. 1964–74, 1985. *International Financial Statistics*. Vols. 17–27 (1964–74); vol. 38, no. 1. (1985) Washington, D.C.: IMF.

INCORA [Instituto Colombiano de Reforma Agraria]. 1966. *Compendio de ingresos y costos de producción de los cultivos transitorios en zonas de crédito supervisado*. Bogotá: INCORA.

———. 1970. *La realidad rural y la reforma agraria como factor de cambio*. Bogotá: INCORA.

———. 1971. *Informe del comité evaluador de la reforma agraria*. Bogotá: Banco Ganadero. Reprinted in part in Arrubla, ed., *Agricultura colombiana*.

Inter-American Committee on the Alliance for Progress. 1968. *Domestic Efforts and the Needs for External Financing for the Development of Colombia*. OEA/ser.H/XIV, CIAP/194. Washington, D.C.: CIAP.

Instituto Geográfico Agustín Codazzi. 1971. *Diccionario geográfico de Colombia*. Bogotá: Editorial Andes.

Johnston, Bruce, and John Mellor. 1961. "The Role of Agriculture in Economic Development." *American Economic Review* 51, no. 4.

Johnston, Bruce, and Peter Kilby. 1975. *Agriculture and Structural Transformation: Economic Strategies in Late-Developing Countries*. New York: Oxford University Press.

Kabeer, Naila. 1985. "Do Women Gain from High Fertility?" In Haleh

Afshar, ed., *Women, Work, and Ideology in the Third World*. London: Tavistock.

Kalmánovitz, Salomón. 1974. "La agricultura en Colombia, 1950–1972." *Boletín Mensual de Estadística*, nos. 276–78. DANE: Seminario Permanente de Problemas Colombianos, Bogotá.

———. 1978. *Desarrollo de la agricultura en Colombia*. Bogotá: Editorial La Carreta.

Kautsky, Karl. 1902. *Die Agrarfrage: Eine Ubersicht die Tendenzen der Modernen Landwirthschaft und die Agrarpolitik der Sozialdemokratie*. Stuttgart: J. H. W. Dietz. Parts of this work appear in English, translated and summarized by J. Banaji in *Economy and Society* 5, no. 1 (1976). Available in Spanish as *La cuestión agraria*. Paris: Editorial Ruedo-Iberico, 1970.

Kelly, Kevin D. 1979. "The Independent Mode of Production." *Review of Radical Political Economics* 11, no. 1.

Kroeber, A. L. 1948. *Anthropology*. New York: Harcourt, Brace.

Kuznesof, Elizabeth. 1980. "An Analysis of Household Composition and Headship as Related to Changes in Mode of Production: São Paulo, 1765 to 1836." *Comparative Studies in Society and History* 22, no. 1.

Kynch, Jocelyn, and Amartya Sen. 1983. "Indian Women: Well-being and Survival." *Cambridge Journal of Economics* 7, no. 314.

*Latin American Perspectives*. 1978–80. Nos. 18 (Summer 1978), 19 (Fall 1978), 27 (Fall 1980).

*Latin American Regional Report, Andean Group*. 1982. RA-82-03 (2 April 1982). London: Latin American Newsletters Ltd.

Leaño Castellanos, Edgard. 1973. "Estudio del uso actual y tenencia de la tierra y socio-económico de la cuenca hidrográfico del Río Yumbo." Manuscript. Palmira: Universidad Nacional de Colombia.

LeGrand, Catherine. 1980. "From Public Lands into Private Property: Landholding and Rural Conflict in Colombia, 1870–1936." Ph.D. diss., Stanford University.

Lenin, V. I. 1920. *The Development of Capitalism in Russia*. Moscow: Progress Publishers, 1964.

León de Leal, Magdelena, and Carmen Diana Deere, eds. 1980. *Mujer y capitalismo agrario*. Bogotá: ACEP.

Lloyd-Jones, Donald. 1961. "The Potential Economic Development of the Upper Cauca Valley, Colombia." Ph.D. diss., Columbia University.

Londoño, B. Rocío, and C. Fabio Velásquez. 1974. *Proceso económico y jurídico-político de Colombia*. Bogotá: Ediciones Camilo.

Long, Norman. 1984. Introduction to Norman Long, ed., *Family and Work in Rural Societies: Perspectives on Non-Wage Labour*. London: Tavistock.

López C., Hugo. 1975. "La inflación en Colombia en la década de los veinte." *Cuadernos Colombianos*, no. 5.

López Cordovez, Luis. 1982. "Agricultura y alimentación: Evolución y transformaciones mas recientes en América Latina." *Revista de la Cepal*, no. 16.

López Toro, Álvaro. 1970. *Migración y cambio social en Antioquia durante el siglo XIX*. Bogotá: Universidad de los Andes.

McGreevey, William Paul. 1970. "Exportaciones y precios de tabaco y cafe." In Urrutia and Arrubla, eds., *Compendio de estadísticas*.

———. 1971. *An Economic History of Colombia, 1845–1930*. Cambridge: Cambridge University Press.

———. 1974. "Urban Growth in Colombia." *Journal of Interamerican Studies and World Affairs* 16, no. 4.

Machado C., Absalón. 1977. *El café: De la aparcería al capitalismo*. Bogotá: Punta de Lanza.

———. 1981. "Políticas Agrarias en Colombia." In Fajardo et al., *Campesinado y capitalismo*.

McNamara, Robert. 1973. *Address to the Board of Governors*. Nairobi: World Bank.

Mallon, Florencia. 1983. *The Defense of Community in Peru's Central Highlands: Peasant Struggle and Capitalist Transition, 1860–1940*. Princeton, N.J.: Princeton University Press.

Mann, Susan A., and James M. Dickinson. 1978. "Obstacles to the Development of Capitalist Agriculture." *Journal of Peasant Studies* 5, no. 4.

Martin, Everett G. 1979. "Colombia Slows Growth of Coffee Output to Prevent World Glut and Price Decline." *Wall Street Journal*, 18 September.

Marx, Karl. 1857. *Grundrisse der Kritik der Politischen Ökonomie (Rohentwurf)*. Available in English as *Pre-Capitalist Economic Formations*. New York: International Publishers, 1965.

Medrano S., Diana. 1980. "La mujer en la región cafetera del suroeste antioqueño." In Léon de Leal and Deere, eds., *Mujer y capitalismo*.

Meillassoux, Claude. 1981. *Maidens, Meal, and Money: Capitalism and the Domestic Community*. Cambridge: Cambridge University Press.

Mejía, Álvaro Tirado. 1971. *Introducción a la historia económica de Colombia*. Bogotá: Universidad Nacional.

———. 1978. "Colombia: Siglo y medio de bipartidismo." In Arrubla et al., *Colombia*.

Melo, Jorge Orlando. 1978. "La república conservador." In Arrubla et al., *Colombia*.

Mill, John Stuart. 1848. *Principles of Political Economy*. Boston: C. D. Little and J. Brown.

MINAGRICULTURA [Ministerio de Agricultura]. 1979. *Cifras del sector agropecuario 1977*. Bogotá: MINAGRICULTURA.

Misas, Gabriel. 1973. "Contribución al estudio del grado de concentración en la industria colombiana." *Boletín Mensual de Estadística* (DANE), no. 266.

Mitchell, Juliet. 1974. *Psychoanalysis and Feminism*. London: Allen Lane.

Molyneux, Maxine. 1979. "Beyond the Domestic Labour Debate." *New Left Review*, no. 116.

Moncayo, V. Manuel, and Fernando Rojas. 1979. "Producción campesina y capitalismo." Bogotá: CINEP [Centro de Investigación y Educación Popular].

Moore, Barrington, Jr. 1969. *Social Origins of Dictatorship and Democracy: Lord and Peasant in the Making of the Modern World*. London: Penguin Books.

Mosk, Carl. 1983. *Patriarchy and Fertility: The Evolution of Natality in Japan and Sweden, 1880–1960*. New York: Academic Press.

Motta de Correa, Lilián. 1980. "Transformación de la unidad doméstica y trabajo de la mujer campesina en una zona de avanzado desarrollo capitalista." In León de Leal and Deere, eds, *Mujer y capitalismo*.

Mouzelis, N. 1976. "Capitalism and the Development of Agriculture." *Journal of Peasant Studies* 3, no. 4.

Mueller, Eva. 1976. "The Economic Value of Children in Peasant Agriculture." In G. Ridker, ed., *Population and Development*. Baltimore: Johns Hopkins University Press.

NACLA [North American Congress on Latin America]. 1978. "Agribusiness Targets Latin America." *NACLA Reports* 12, no. 1.

Nag, Moni. 1972. "Sex, Culture, and Human Fertility: India and the United States." *Current Anthropology* 13 (April).

Nag, Moni, Benjamin White, and Robert Peet. 1977. "Economic Value of Children in Two Peasant Societies." In *International Population Conference: Mexico, 1977*, vol. 1. Liège, Belgium: Union for the Scientific Study of Population.

Nariño, Departamento de. 1979. *Nariño 1977: Anuario estadístico, plan departamental*. Pasto: Departamento de Nariño.

Nash, Manning. 1966. *Primitive and Peasant Economic Systems*. San Francisco: Chandler.

Nerlove, Marc. 1974. "Household and Economy: Toward a New Theory of Population and Economic Growth." *Journal of Political Economy* 82, no. 2, part 2.

Nieto Arteta, Luís Eduardo. 1975. *El café en la sociedad colombiana*. Bogotá: Tiempo Presente.

Nikolitch, Radoje. 1969. "Family-Operated Farms: Their Compatibility with Technical Advance." *American Journal of Agricultural Economics* 51, no. 3.

Ochoa, Luis Hernando. 1982. "El descenso de la fecundidad en Colombia y sus implicaciones demográficas." In L. H. Ochoa, ed., *Implicaciones socio-económicas y demográficas del descenso de la fecundidad en Colombia*. Monografías de la CCRP, vol. 18. Bogotá: CCRP.

Ortega, Alfredo. 1920. *Ferrocarriles colombianos: Resumen histórico*. 2 vols. Bogotá: Biblioteca de Historia Nacional.

Ortega, Emiliano. 1982. "La agricultura campesina en América Latina: Situaciones y tendencias." *Revista de la Cepal*, no. 16.

Ortiz, Sutti. 1973. *Uncertainties in Peasant Farming: A Colombian Case*. London: Athlone Press.

———. 1984. "Colonization in the Colombian Amazon." In Marianne Schmink and Charles H. Wood, eds., *Frontier Expansion in Amazonia*. Gainesville: University of Florida Press.

Ospina Vásquez, Luis. 1955. *Industria y protección en Colombia, 1810–1930*. Medellín: Editorial Santa Fe.

Owen, Wynn F. 1966. "The Double Development Squeeze on Agriculture." *American Economic Review* 56, no. 1.

Pacheco, Margarita Rosa. 1980. "Ejidos de Cali: Siglo XIX." *Historia y Espacio: Revista de Estudios Históricos Regionales* (Colombia) 2, nos. 6–7.

Palacios, Marco. 1980. *Coffee in Colombia, 1850–1970: An Economic, Social, and Political History*. New York: Cambridge University Press.

Parsons, Donald O. 1984. "On the Economics of Intergenerational Control." *Population and Development Review* 10, no. 1.

Parsons, James L. 1968. *Antioqueño Colonization in Western Colombia*. Berkeley and Los Angeles: University of California Press.

Patnaik, Utsa. 1979. "Neo-Populism and Marxism: The Chayanovian View of the Agrarian Question and its Fundamental Fallacy." *Journal of Peasant Studies* 6, no. 4.

Payer, Cheryl. 1980. "The World Bank and the Small Farmer." *Monthly Review* 33, no. 6.

Pearse, Andrew, and Salomón Rivera. 1963. *La tenencia de la tierra y sus implicaciones socio-económicos en Tenza, Colombia: Un estudio de minifundio*. Bogotá: Universidad Nacional.

Pérez Zapata, Hernán. 1978. *En juiciamento a la política agraria y cafetera*. Bogotá: Asociación Colombiana de Ingenieros Agrónomos.

Piñeda Giraldo, Roberto. 1960. *El impacto de la Violencia en el Tolima: El caso de el Líbano*. Monografías Sociológicas, no. 6. Bogotá: Universidad Nacional.

Posada, Antonio, and Jeanne Posada. 1966. *The CVC: Challenge to Underdevelopment and Traditionalism.* Bogotá: Ediciones Tercer Mundo.

Potter, Jack, George Foster, and May Diaz, eds. 1967. *Peasant Society: A Reader.* Boston: Little, Brown.

PRODESARROLLO [Federación Nacional de Cafeteros, Programa de Desarrollo y Diversificación para las Zonas Cafeteras]. 1968. *Substitución zonas cafeteras marginales.* Manizales: PRODESARROLLO.

———. 1972a. *Estudio de zonificación y uso potencial del suelo de Valle del Cauca.* Bogotá: PRODESARROLLO.

———. 1972b. *Proyecto de diversificación agropecuaria para la zona cafetera central de Colombia.* Bogotá: PRODESARROLLO.

Rao, C. H. H. 1971. "Uncertainty, Entrepreneurship, and Sharecropping in India." *Journal of Political Economy* 79, no. 3.

Raup, Phillip. 1973. "Corporate Farming in the United States." *Journal of Economic History* 33, no. 1.

Redclift, Michael. 1983. "Production Programs for Small Farmers: Plan Puebla as Myth and Reality." *Economic Development and Cultural Change* 31, no. 3.

Reinhardt, Nola. 1981. "The Independent Family Farm Mode of Production: El Palmar, Colombia, 1890–1978, A Study of Economic Development and Agricultural Change." Ph.D. diss., University of California, Berkeley. Ann Arbor, Mich.: University Microfilms.

———. 1983. "Commercialization of Agriculture and Rural Living Standards: El Palmar, Colombia, 1960–1979." *Journal of Economic History* 43, no. 1.

———. 1986. "The Consolidation of the Import-Export Economy in Nineteenth-Century Colombia: A Political-Economic Analysis." *Latin American Perspectives* 13, no. 1.

———. 1987a. "Economic Development and Rural Fertility in Theory and Practice: Evidence of Change from El Palmar, Colombia." Paper presented to the Gender Issues in International Development Seminar, Smith College, Northampton, Mass., 16 March.

———. 1987b. "Modernizing Peasant Agriculture: Lessons from El Palmar, Colombia." *World Development* 15, no. 2.

Rey, Pierre Philippe. 1973. *Les Alliances des classes.* Paris: Maspero.

Rippy, J. Fred. 1970. *El capital norteamericano y la penetración imperialista en Colombia.* Medellín: Editorial La Oveja Negra.

———. 1976. *The Capitalists and Colombia.* New York: Arno Press.

Rochín, Refugio I., and Diego Londoño R. 1975. "Structural Characteristics of Garcia Rovira, Colombia: Problems for Integrated Rural Development." Paper presented at the Rural Development Workshop, University of California, Berkeley.

Rodríguez R., Jorge E., and William P. McGreevey. 1970. "Colombia: Comercio exterior, 1835–1962." In Urrutia and Arrubla, eds., *Compendio de estadísticas.*

Roldán Luna, Diego. 1979. "La propiedad parcelaria campesina." Ph.D. diss., Sussex University, England.

Roseberry, William. 1983. *Coffee and Capitalism in the Venezuelan Andes.* Austin: University of Texas Press.

Rosenzweig, Marc R. 1977. "The Demand for Children in Farm Households." *Journal of Political Economy* 85, no. 1.

Rubin, Gayle. 1975. "The Traffic in Women: Notes on the Political Economy of Sex." In Rayna Reiter, ed., *Toward an Anthropology of Women.* New York: Monthly Review Press.

Ruddle, Kenneth, and Kathleen Barrows. 1974. *The Statistical Abstract of Latin America, 1972.* Los Angeles: University of California.

Ruiz, Soledad. 1972. "La fuerza de trabajo en la zona cafetera de Tolima." Internal DANE document. Bogotá: DANE.

Safford, Frank. 1966. *Commerce and Enterprise in Central Colombia, 1821–1870.* Diss. reprint. Ann Arbor, Mich.: University Microfilms.

Salazar, Oscar Rodríguez. 1973. *Efectos de la grán depresión sobre la industria colombiana.* Bogotá: Ediciones El Tigre de Papel.

Sandilands, Roger. 1971. "Modernización del sector agropecuaria." *Revista de Planeación y Desarrollo* (Colombia) 3 (October).

Sandoval, Rodrigo Parra. 1966. *La estructura social y el cambio en la tecnología agrícola: El caso de Candelaria.* Monografías Sociológicas, no. 21. Bogotá: Universidad Nacional.

Schejtman, Alexander. 1980. "The Peasant Economy: Internal Logic, Articulation, and Persistence." *CEPAL Review*, no. 11.

Schultz, Theodore. 1964. *Transforming Traditional Agriculture.* New Haven: Yale University Press.

———. 1965. *Economic Crisis in World Agriculture.* Ann Arbor: University of Michigan Press.

———, ed. 1974. *Economics of the Family.* Chicago: University of Chicago Press.

Schultz, T. Paul. 1981. *Economics of Population.* Reading, Mass.: Addison-Wesley.

Schwartz, Stuart B. 1985. *Sugar Plantations in the Formation of Brazilian Society: Bahia, 1550–1835.* Cambridge: Cambridge University Press.

Science for the People. 1979. *Special Issue on Food, Agriculture and Agribusiness* 11, no. 3.

Seccombe, Wally. 1974. "The Housewife and Her Labour under Capitalism." *New Left Review*, no. 83.

Servolin, Claude. 1972. "L'Agriculture dans le mode de production capitaliste." In *L'Univers politique des paysans*. Paris: Fondation National des Sciences Politiques.

Shanin, Teodor. 1971. *Peasants and Peasant Societies*. Harmondsworth, England: Penguin Books.

———. 1972. *The Awkward Class: Political Sociology of the Peasantry in a Developing Society, Russia 1910–1925*. London: Oxford University Press.

———. 1974. "The Nature and Logic of the Peasant Economy." *Journal of Peasant Studies* 1, nos. 1–2.

Silva Holguin, Raul. 1960. *Valle del Cauca, tierra del promesión*. Cali: Imprenta Departamental.

Smith, Carol. 1984a. "Forms of Production in Practice: Fresh Approaches to Petty Commodity Production." *Journal of Peasant Studies* 11, no. 4.

———. 1984b. "Labor and International Capital in the Making of a Peripheral Social Formation: Economic Transformations in Guatemala, 1850–1980." In Charles Bergquist, ed., *Labor in the Capitalist World-Economy*. Beverly Hills, Calif.: Sage.

Smith, T. Lynn. 1944. *Tabío: Estudio de la organización social rural*. Bogotá: Minerva. Available in English as *Tabio: A Study in Rural Social Organization*. Washington, D.C.: USDA, 1945.

Standing, Guy. 1978. *Labour Force Participation and Development*. Geneva: International Labour Office.

———. 1983. "Women's Work Activity and Fertility." In R. A. Bulatao and R. D. Lee, eds., *The Determinants of Fertility in Developing Countries*, vol. 1. New York: Academic Press.

Standing, Hilary. 1985. "Resources, Wages, and Power: The Impact of Women's Employment on the Urban Bengali Household." In Haleh Afshar, ed., *Women, Work, and Ideology in the Third World*. London: Tavistock.

Stavenhagen, Rodolfo, ed. 1970. *Agrarian Problems and Peasant Movements in Latin America*. New York: Anchor Books.

Stichter, Sharon. 1975–76. "Women and the Labor Force in Kenya, 1895–1964." *Rural Africana* 29.

———. 1985. *Migrant Laborers*. Cambridge: Cambridge University Press.

Stichter, Sharon, and Jane Parpart. 1987. "Introduction: Towards a Materialist Perspective on African Women." In Stichter and Parpart, eds., *Patriarchy and Class: African Women in the Home and Workforce*. Boulder, Colo.: Westview Press.

Stys, W. 1957. "The Influence of Economic Conditions on the Fertility of Peasant Women." *Population Studies* 11, no. 2.

Tanker, Irene, and Michele Bramsen, eds. 1976. *Women and World Development*. Washington, D.C.: Overseas Development Council.

Taussig, Michael. 1974. "Rural Proletarianization: A Social and Historical Enquiry into the Commercialization of the Southern Cauca Valley, Colombia." Ph.D. diss., University of London.

Tax, Sol. 1953. *Penny Capitalism: A Guatemalan Indian Community*. Washington, D.C.: Smithsonian Institute of Social Anthropology.

Thompson, E. P. 1963. *The Making of the English Working Class*. New York: Vintage Books.

Thorner, Daniel. 1966. "Chayanov's Concept of Peasant Economy." In D. Thorner, B. Kerblay and R. E. F. Smith, eds., *The Theory of Peasant Economy*. Homewood, Ill.: Richard D. Irwin.

Tobón C., José Hiriam. 1981. "Experiencias en cambios tecnológicos agrícolas y su transferencia en el Proyecto Oriente Antioqueño de Colombia." In Andrés R., Novóa B., and Joshua L. Posner, eds., *Agricultura de ladera en América tropical*. Costa Rica: CATIE.

Todaro, Michael P. 1985. *Economic Development in the Third World*. 3d ed. New York: Longman.

Torres, Hugo A. 1971. "Analysis of the Structure, Conduct, and Performance of Grain Markets in the Cauca Valley, Colombia." In Kenneth L. Turk, ed., *Some Issues Emerging from Recent Breakthroughs in Food Production*. Ithaca, N.Y.: New York State College of Agriculture.

Tovar, Hermes. 1975. *El movimiento campesino en Colombia durante los siglos XIX y XX*. Bogotá: Ediciones Libres.

United Nations. 1949. *1948 Demographic Yearbook*. New York: United Nations.

———. 1955. *1955 Demographic Yearbook*. New York: United Nations.

———. 1972. *1971 Demographic Yearbook*. New York: United Nations.

———. 1980. *1979 Demographic Yearbook*. New York: United Nations.

UNDP [United Nations Development Program]. 1980. *Rural Women's Participation in Development*. New York: UNDP.

Uribe, Mariya, and Augusto Donoso. 1978. *Evolución de los proyectos de desarrollo rural en Colombia*. Bogotá: Instituto InterAmericano de Ciencias Agrícolas (IICA-OEA).

Urrutia M., Miguel. 1970. "Estadísticas de Precios: 1846–1933." In Urrutia and Arrubla, eds., *Compendio de estadísticas*.

Urrutia M., Miguel, and Mario Arrubla, eds. 1970. *Compendio de estadísticas históricas de Colombia*. Bogotá: Universidad Nacional.

USDA [United States Department of Agriculture]. 1969. *Changes in Agricultural Production and Technology in Colombia*. Foreign Agriculture Economic Report no. 52. Washington, D.C.: USDA.

———. 1970. *Agricultural Productivity in Colombia.* Foreign Agriculture Economic Report no. 66. Washington, D.C.: USDA.

Valencia González, Sofía. 1982. "Experiencias de la Fase I del Programa DRI y recomendaciones para la Fase II." Working paper DNP-UEA-010. Bogotá: DNP.

Valle del Cauca, Departamento de. 1961. *Censo agropecuario del Valle del Cauca, 1959.* Cali: Universidad del Valle.

Vallejo Murillo, Jorge. N.d. "La fuerza de trabajo en el desarrollo de la agricultura capitalista." Mimeographed. Lecture no. 529. Cali: Universidad del Valle, División de Ciencias Sociales y Económicas.

Vallejo Murillo, Jorge, and Sofía Valencia. 1973. *Crédito institucional del sector agropecuario.* Bogotá: OPSA, Ministerio de Agricultura.

Velez, Hugo. 1979. "Producción campesina e inflación en la década de 1970." *Cuadernos Colombianos* 12, no. 3.

Vergara y Velasco, F. J. 1974. *Nueva geografía de Colombia.* 3 vols. 2d ed. Bogotá: Banco de la República.

Vergopoulos, Kostas. 1975. *The Agrarian Problem in Greece: The Issue of the Social Incorporation of Agriculture.* In Greek. Athens: Exantas.

———. 1978. "Capitalism and Peasant Productivity." *Journal of Peasant Studies* 5, no. 4.

Villegas, Jorge. 1977. *Colombia: Colonización de vertiente en el siglo 19.* Medellín: Centro de Investigaciones Económicas, Universidad de Antioquia.

Warman, Arturo. 1980. *Ensayos sobre el campesinado en México.* Mexico City: Editorial Nueva Imagen.

Wiegersma, Nancy. 1985. "The Rise of Patriarchy and the Destruction of the Collective in Vietnam." Paper presented at the Allied Social Science meetings, New York, December.

Willis, R. J. 1974. "Economic Theory of Fertility Behavior." In Theodore Schultz, ed., *Economics of the Family.*

Wolf, Eric. 1966. *Peasants.* Englewood Cliffs, N.J.: Prentice-Hall.

Wolpe, Harold. 1972. "Capitalism and Cheap Labour Power in South Africa: From Segregation to Apartheid." *Economy and Society* 1, no. 4.

World Bank. 1972. *Economic Growth of Colombia: Problems and Prospects.* Baltimore: Johns Hopkins University Press.

———. 1975. *Rural Development: Sector Policy Paper.* Baltimore: Johns Hopkins University Press.

———. 1979. *World Development Report, 1979.* Washington, D.C.: World Bank.

Yudelman, Montague. 1976. "Agriculture in Integrated Rural Development: The Experience of the World Bank." *Food Policy* (November).

Zambrano, Hernán. 1986. "Tendencias de la caficultura colombiana." *Economía Colombiana,* no. 179.

Zamosc, León. 1979. "Notas teóricas sobre la subordinación campesina al capital." *Estudios Rurales Latinoamericanos* 2, no. 3.

Zuleta, Estanislao. 1973. "Proceso evolutivo de la propiedad." In *La tierra en Colombia.* Cuadernos La Oveja Negra, no. 6. Medellín: Editorial La Oveja Negra.

Zuleta, Luís Alberto. 1975. "El sector cafetero y los fenómenos inflacionarios." *Cuadernos Colombianos,* no. 7.

# Index

Compositor: G&S Typesetters, Inc.
Text: 11/13 Caledonia
Display: Caledonia
Printer: Braun-Brumfield, Inc.
Binder: Braun-Brumfield, Inc.